Forest and Plant Bioproducts Division 2014

Core Programming Area at the 2014 AIChE Annual Meeting

Atlanta, Georgia, USA
16-21 November 2014

ISBN: 978-1-5108-1260-4

Printed from e-media with permission by:

Curran Associates, Inc.
57 Morehouse Lane
Red Hook, NY 12571

Some format issues inherent in the e-media version may also appear in this print version.

Printed by Curran Associates, Inc. (2015)

AIChE
120 Wall Street, FL 23
New York, NY 10005-4020

Phone: (800) 242-4363
Fax: (203) 775-5177

www.aiche.org

Additional copies of this publication are available from:

Curran Associates, Inc.
57 Morehouse Lane
Red Hook, NY 12571 USA
Phone: 845-758-0400
Fax: 845-758-2634
Email: curran@proceedings.com
Web: www.proceedings.com

TABLE OF CONTENTS

392737 Innovations in Forest Products Industry and the role of a Chemical Engineer

Abstract:

Forest Product s Industry is a major industry in North America employing a large number of Chemical Engineers where they play important roles. One such important role is in the conception/development/implementation of New/Improved Products and Processes. The author will present a series of recent innovations (processes/products) and analyze/identify additional skill requirements for a chemical engineer and would conclude that a major need exits in proficiency in creativity and innovation tools to accelerate the process of innovation. The presenter will elaborate on the techniques of creative problem solving further emphasizing their importance in successfully taking advantage of major emerging growth opportunities for the industry.

392738 Characterization, Fractionation, Separation and Hydro ysate Toxicity Reduction on Fermentation Of C5 and C6 Sugars Derived From Lignocellulosic Biomass

Abstract:

Biomass pretreatment is necessary to reduce the calcitrance of lignocellulosic feedstocks to enzyme saccharification and to allow the liberation of soluble sugars (C5 and C6) from the plant cell wall. However, pretreatment can also release undesirable compounds into the hydrolysate that can inhibit the growth and product yields of fermentation organisms. It is critical to understand and characterize the toxicity effects of fermentation hydrolysate to devise optimal pretreatment and down stream inhibitor sequestering process, that maximizes cellulosic sugar recovery, conversion and enhance productivity of the test organisms .

Sweetwater Energy have a proprietary modular pretreatment system for extracting sugars from biomass. We conducted systematic study to characterize the key inhibitors and toxic molecules in the hydrolysate that the necessary parameters to reduce and eliminate these inhibitors around hydrolysate toxicity in various the dilute acid pretreated feedstocks . Initial results and the outline of the strategy directed toward further refining the cellulosic sugars using membranes and ion exchange resins and its combination would be presented. In addition, the problems and the prospects of establishing a cost effect integrated down stream separation technology that renders the cellulosic sugar mixtures of C5 and C6 syrups and make them emendable to the next stage of fermentation will be highlighted.

Pyrolysis Kinetics of Deinking Sludge

Shilpa Kulkarni [a 1], Vivek Kumar [b 1], Sanjeev K. Singh[2]., Mukesh C. Bansal [3]

[1]Department of Paper Technology, Indian Institute of Technology- Roorkee, India,
[2]CSIR- Central Building Research Institute, Roorkee, India,
[3]Bharat Institute of Technology-Meerut, India
[a]shilpa.ips@gmail.com [b]vivekfpt@gmail.com [2]sksingh_cbri@yahoo.co.in

---

Abstract

Deinking sludge is a solid waste generates from the recycled paper industry. During last decades, disposal of deinking sludge in landfills was a very popular option, however it creates serious environmental problem. To overcome this problem, alternate disposal method can be to utilize deinking sludge for fuel generation through thermo chemical conversion process techniques. In the present work kinetics of deinking sludge pyrolysis has been studied using TGA. The weight loss process of deinking sludge with change of temperature was recorded. The results revealed that the weight loss can be divided into four distinct zones: dewater stage, volatile releasing, burnout carbon and decomposition of some calcium carbonate. These distinct zones can be determined by thermogravimetry analyzer (TGA). The effect of heating rate 10 C min^{-1} and sample weight (10 mg) of deinking sludge has been determined in the presence of nitrogen atmosphere having 200 ml/min flow rate. The temperature ranges from (ambient to 900°C) was used throughout the experiments.

Keywords: Recycled Paper Sludge; Pyrolysis; Thermal Degradation

1. Introduction

Recycled paper industry generates significant amount of solid sludge, generally known as paper industry sludge. The sludge of recycled paper industry contains fillers, fibers, ink and heavy metals [1]. If solid waste is not handled properly, it can create serious environmental problems. Therefore, handling of the recycled paper sludge is the need of the hour. Till now conventional methods were of disposing sludge in agricultural land, dumping in ocean, including landfilling but day by day these techniques are losing their popularity which bounds us to explore new ways of sludge disposal [2]. In a recent review, Norris [3] suggested that waste minimization must be a top priority in paper manufacturing industries. Nussbaumer [4] discovered combustion for managing solid waste accompanied by energy recovery. Moreover, alternate route can be through pyrolysis.

Pyrolysis is a process by which organic substances are thermally decomposed in an oxygen-free atmosphere at temperatures up to 900° C which produces different products such as bio-gas, bio-oils, charcoal products etc. These products can be wisely applied in variety of fields. Pyrolysis of wood and other biomaterials is also utilized to produce chemicals. In the present scenario, more and more researchers are following the path of pyrolysis [5]. Moreover, inspite of the fact that pyrolysis of variety of materials has been done till today but still limited information is available on pyrolysis of deinking sludge.

Present paper aims is study the Pyrolysis of deinking sludge and analysis of their by-product (bio-oil and bio-char). The thermal degradation study of sludge was also studied.

2. Material and methods

2.1 *Raw material*

Deinking sludge was selected as raw material for the production of pyrolysis by-products. Sludge sample was collected from dewatering unit of primary effluent from flotation cell. Preparation of raw material for pyrolysis was done by drying sludge in sunlight for 3 days and then oven dried at 50°C for 48 hrs. Dried deinking sludge (DPS) sample was ground. Ground sludge sample was further used for pyrolysis.

2.2 *Thermal analysis of deinking sludge*

The thermal degradation characteristic of deinking paper sludge in nitrogen atmosphere at 10 °C min^{-1} heating rate was studied. The TGA and DTG curves are shown in Figure 1 indicate that degradation process happens in three stages in nitrogen atmospheres besides moisture removal. The onset temperature of stage one , two and three (in nitrogen atmosphere) were found to be ~155°C , ~ 376 °C and ~ 555°C at heating rate 10 °C min^{-1} respectively. The results revealed that the weight-loss process of deinking sludge was a non-isothermal reaction and composed of four stages, i.e. dewater stage, volatile releasing stage, carbon burnout stage and some calcium carbonate decomposition

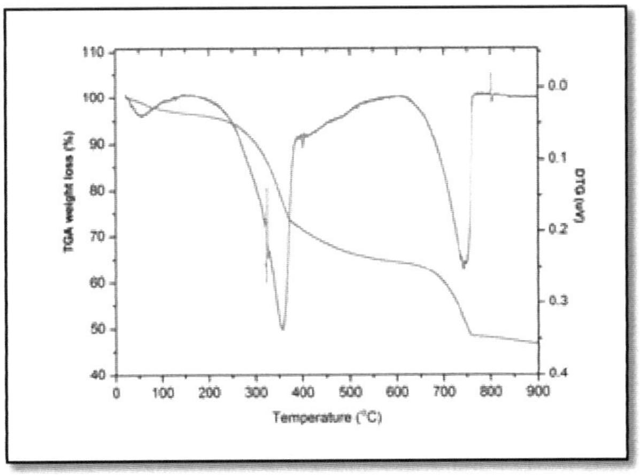

Figure 1: TGA and DTG curves at 10°C/min heating rates

Table 1: Thermal degradation of deinking sludge in different zones in the presence of Nitrogen atmosphere at 200 ml/min flow rate

Reaction zones	Heating rates (°C min^{-1})	Water evolved (%)	Degradation temperature (°C)			Total degradation (wt. %)	Degradation Rates (wt.% min^{-1})	Residue at 900°C (wt. %)
			Initial	Final	Maximum			
I	10	2.5	~ 155	~ 376	~ 330	22.3	3.7	-
II	10	-	~ 376	~ 555	-	13.0	-	-
III	10	-	~ 555	~ 850	~ 750	18.4	0.9	43.8

2.3 *Characterization of pyrolysis products*

Standard Test Methods for Proximate Analysis of the Analysis Sample of Coal and Coke by Instrumental Procedures1 (D 5142 – 02a) was used for determination of bio-char. Thermo gravimetric analyzer (TGA) (ESTAR TG/DTG 6300 SII) was used for thermal degradation study of raw deinking sludge sample in which the weight loss with respect to temperature was determined. The temperature ranges from 25℃ to 800℃ with the heating rate of 10℃ min^{-1} under a flow of Nitrogen at 200 ml min^{-1} were used. Bio-oil was analysis by using HPLC (High Pressure Liquid Chromatography).

2.4 *Pyrolysis Process set-up and procedure*

Pyrolysis was carried out in slow pyrolysis reactor heated of type fixed bed up flow reactor furnace. At the beginning of each run, 20 g of pre-treated deinking sludge was placed in the glass reactor tube made of quartz. Highpurity (99.995%) nitrogen was used as the purging gas, flowing through the sample bed at 50-100 ml/min. The reactor was heated to the designated temperature from ambient to 500 ℃ at 10 ℃/min heating rate. The pyrolysis equipment was set up and is shown in **Figure 2**. The pyrolysis temperature varied from 180℃, 300℃, and 500℃ respectively. After the pyrolysis process, the reactor was cool down to the room temperature before the pyrolysis residue was removed for quantification and characterization. Process parameters for pyrolysis of deinking sludge were fixed throughout the experiment which is shown in **Table 2.**

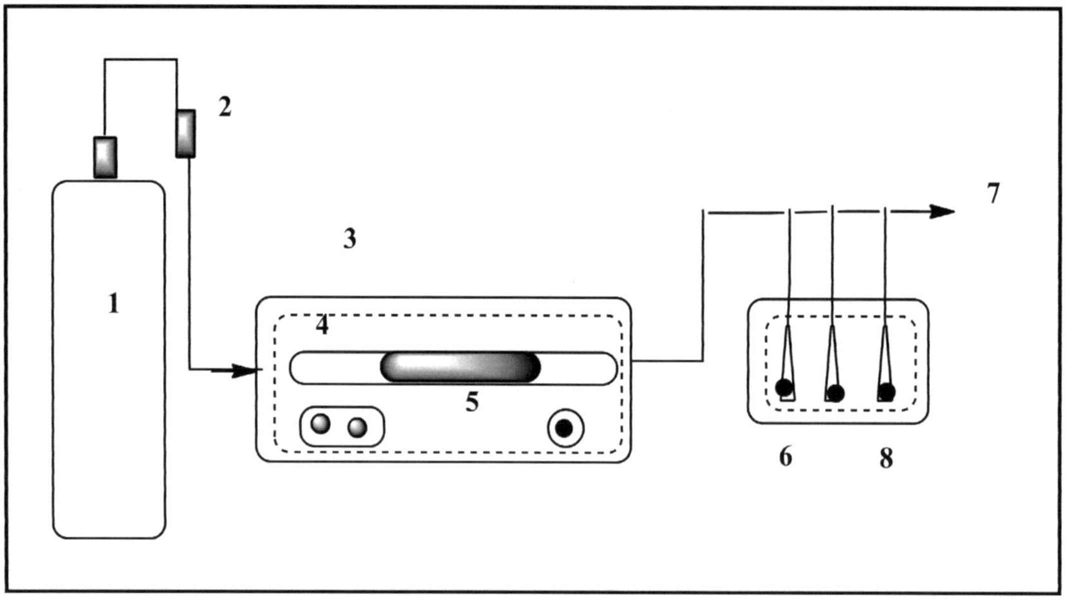

Figure 2: Experimental set-up for pyrolysis of deinking sludge. Following accessories were used for performing experiments these are: Nitrogen cylinder (1); Mass flow (2); Furnace (3); Glass reactor tube (4), bio-char (5); condenser (6); bio-gas collect (7) and bio-oil collect (8)

Table 2: Process Parameters for Pyrolysis of deinking sludge	
Heating rate (°C/min)	10°C/min
Inert atmosphere	Pure Nitrogen
Flow of nitrogen	50-100 ml/min
Type of pyrolysis reactor	Fixed bed up flow
Reactor	Slow pyrolysis

3. Results and discussion

Pyrolysis temperature is the key factor influencing the biochar yield. To study the effect of temperature, biochar yield, sludge was pyrolyzed at different temperatures (180-500 °C) by maintaining the heating rate 10 °C/min with nitrogen 50-100 L/min. The pyrolysis time was kept constant (i.e. 2 hrs) throughout the pyrolysis experiments. It was observed that biochar yield decreased with increase in pyrolysis temperature and maximum yield was observed at 300°C which shown in Table 4. Proximate analysis of bio-char of deinking sludge was shown in Table 3. The major chemical composition of bio-oil shown in Table 5. Bio-oil analysis was carried out by using HPLC (High Pressure Liquid Chromatography).

High performance liquid chromatography (HPLC) is a chromatographic technique used to separate a mixture of compounds in analytical chemistry and biochemistry with the purpose of identifying, quantifying or purifying the individual components of the mixture. Separation is based on the analyses' relative solubility between two liquid phases one is mobile phase (solvent) and other is stationary phase (Bonded phase). For this analysis C-18 (Octadecyl Silyl) column is used as a stationary phase while for mobile phase dichloromethane. Flow rate 0.5 ml/min was used. After collection of bio-oil from pyrolysis, it can further prepared for analysis. Take bio-oil (1-2 ml) sample mix with dichloromethane solvent for analysis. Dichloromethane procured from Hi-media, 99.9% pure was used.

Table 3: Proximate analysis of bio-char of deinking sludge on dry basis	
Moisture content (%)	1.6
Volatile matter (%)	31.16
Ash content (%)	61.19
Fixed carbon (by difference, %)	4.45

Table 4: %Yield of Pyrolysis products of deinking sludge on dry basis

	Bio-char (wt. %)	Bio-oil (wt. %)
At 180°C	76.6	11.6
At 300°C	93.1	6.68
At 500°C	66.17	3.43

Table 5: Major chemical composition of bio-oil		
No. Peaks	Retention time (minutes)	Area
1	0.724	9807
2	3.576	7871
3	5.240	8815
4	5.470	7007
5	7.968	8280
6	8.173	1937
7	12.702	3119
8	12.843	2904
1	14.639	3414
2	14.791	6167
1	2.293	1998
2	2.367	1213
3	6.859	6939
4	7.009	5779
5	7.464	11619
6	7.628	9552
7	8.062	13100
8	8.352	5252
9	8.544	1676
10	8.785	3737
11	11.979	1629
12	16.403	18400
13	16.862	8243
14	18.140	2016

4. Conclusions

The following conclusion can be drawn from above experiments:

➢ TGA and DTG curves showed a number of distinct reaction zones during the thermal degradation process for deinking sludge. Natural break in the slope of TGA is characterized by the change in the rate of weight loss due to drying, volatilization and degradation.

➢ Maximum % yield of char was obtained at 300℃ while maximum % yield of liquid was obtained at 180℃

➢ Experiments needs more investigation in that area like to optimize process parameters, analysis through GC/MS and bio-gas analysis.

➢ Study on Pyrolysis experiments beyond 500℃ with different types of reactors will require

Acknowledgement

Authors sincerely thanks to Recycled paper industry, Punjab for providing us deinking sludge for experimental purpose and also thanks to Dr T. Bhaskar, Indian Institute of Petroleum-Dehradun for providing lab facility.

References

[1] K. Liu, X.Q. Ma, H.M. Xiao, ***"Experimental and kinetic modeling of oxygen-enriched air combustion of paper mill sludge"***, Journal of Waste Management 30 (2010) 1206-1211.

[2] L. Yanfen, X. Ma, ***"Thermogravimetric analysis of the co-combustion of coal and paper mill sludge"***, Journal of Applied Energy 87 (2010) 3526-3532.

[3] M. Norris, L.W. Titshall, ***"The Potential for Direct Application of Papermill Sludge to Land: A Greenhouse Study"***, International Journal of Environmental Research 5 (2011) 673-680.

[4] T. Nussbaumer, ***"Combustion and Co-combustion of Biomass: Fundamentals, Technologies, and Primary Measures for Emission Reduction"***, Journal of Energy & Fuels 17 (2003) 1510-1521

[5] L. Rui, S. Wu, G. Lv, Q. Yang, ***"Energy and resource utilization of deinking sludge pyrolysis"***, Applied Energy 90 (2012) 46-50.

364319 Slow Dynamics As a Warning Signal for Eruptive Events in Forest Fires

Abstract:

Many complex systems are multistable and can, when sufficiently perturbed, undergo critical transitions in which they shift abruptly between dynamically distinct states (e.g. the onset of climatic shifts, the collapse of ecosystems). Forest fires represent a multistable system for which warning signals for critical transitions could aid in the development of improved fire intervention practices. In forest fires, feedback between wind and spreading flames (i.e. "wind-fire coupling") can spontaneously trigger the formation of blowup fires, large fires with multiplied rates of energy output. Such events, which are not captured by operative models of fire behavior, enhance the unpredictability of spreading fires, pose enormous risks to fire response teams, and complicate fire suppression efforts. To examine the onset of blowup fires in the context of critical transitions, we constructed a physical model for these events based on a bistable combustion system. In this system, flames propagate along contoured strips of nitrocellulose with one of two possible modes of propagation: a structured mode, which burns slowly, and an unstructured mode, which, as a result of wind-fire coupling, burns rapidly. Using this model, we identified conditions where the onset of the second mode was more likely, and by examining dynamics under those conditions, we demonstrated that the system exhibited slower responses to perturbations as it approached a bifurcation point (a phenomenon referred to as "critical slowing down" in dynamical systems theory). This result suggests that, in forest fires, slower responses to terrain, wind, and/or other environmental factors may presage the onset of blowup fires and other dangerous ignition events resulting from wind-fire coupling. With this work, we introduce a simple, experimentally tractable system for examining critical transitions in forest fires and other complex multi-stable systems.

368767 Micro and Nano Scale 3D Structure of Biomaterials and Their Relationship to Material Properties

Abstract:

Biomaterials such as paper, board, bio-based composites and pigmented paper coatings have complex internal structures. Internal structures of porous materials have a significant influence on the efficiency of the manufacturing process, the properties of the materials and their end-use applications. With the advent of novel materials characterization techniques it is now possible to non-intrusively visualize the internal three dimensional structure of porous materials and attempt to better understand structure property functional relationships. X-ray micro computed tomography (X-μCT) is a technique that can be used to non-intrusively visualize micron and sub-micron scale structure of materials. Both absorptive contrast and phase contrast imaging and image reconstruction techniques can be used to obtain the 3D structure. The basic principle of 3D computed tomography is based on capturing the images of the sample by scanning at different angles and reconstructing the three dimensional image using image transformation methodologies.

Using X-ray micro CT three dimensional images of porous biomaterials, we have developed computational methods to characterize the structural features including material porosity, fiber-pore interfacial area, structural tortuosity, 3D pore size distribution, pore-fiber network characteristics such as node density distribution, bond length distribution, and morphological maps of the structure such as pore and bond skeletons. In addition, we have used the actual 3D micro structure of porous materials to predict their liquid and vapor transport properties. Using random walk simulations in 3D structures, we were able to predict the anisotropic simultaneous pore-fiber diffusion characteristics of fibrous porous media. In addition to transport properties, using cell wall mechanical properties and actual 3D structures and a meshed finite element analysis, we continue to develop methods to predict the elastic and viscoelastic mechanical properties of the network structure. Using this approach, effect of fiber properties and processing conditions on material properties can be better discerned.

Even though micron scale structural details may be sufficient in many applications of biomaterials, in order to better understand the ultrastructure of plant cell walls and their role in biomass conversion processes, it is necessary to probe the ultra-structural features at the nanoscale. 3D computed tomography technique mentioned above is now being extended to other imaging techniques such as Transmission Electron Microsopy (TEM). The spatial resolution of the 3D TEM is reported to be of the order of 5 nm. Using TEM nano computed tomography (TEM-nCT), we have been exploring the ultrastructure of biomass cell walls and attempting to characterize their structural features. This technique has immense potential in not only probing the nano scale ultrastructure but also superimposing topo chemical distributions using correlative imaging. These advanced materials characterization tools will enable us to truly "engineer" material structures tailored for specific end-use applications.

383470 Supercritical CO2 Based Pretreatment of Guayule Biomass for High Sugar Yield and Low Inhibition Hydrolysate

Abstract:

Guayule (*Parthenium argentatum* Gray) is a commercial crop for hypoallergenic latex production. Latex (rubber) and resins, however, make up only about 20% of guayule biomass, rest of which, already powderized, is potentially a good lignocellulosic source for biorefinery. Lignocellulosic biorefinery has three common steps: pretreatment, enzymatic hydrolysis and fermentation. In the pretreatment step, the biomass is chemically and/or physically treated to help advance the enzymatic hydrolysis efficiency. In our study supercritical CO_2 based pretreatment was found very effective to achieve higher monomeric sugar yield in enzymatic hydrolysis. This pretreatment involved: raising system temperature, adding water to the bagasse and load them into the reactor, pressurizing using CO_2, holding the system for a period of time, and the quick release of pressure. The supercritical pretreatment preserved essentially all carbohydrate without losing, e.g., hemicellulose sugars as in the widely used dilute acid pretreatment. It had both reaction effect and explosion effect. Because of higher temperature and mildly acidic condition (carbonic acid) in this pretreatment, some fermentation inhibitors such as furfural, hydroxymethyl furfural, and acetic acid could be produced. In this study we optimized the pretreatment conditions for minimal degradation/inhibition products and maximal sugar yield in subsequent enzymatic hydrolysis. It was found that at harsher condition reaction effect dominated without apparent explosion effect. Higher temperature (>180^0C) and pressure (>12.2 MPa) and longer time (>30 min) gave higher sugar yield: as high as 80% of glucose, xylose and total reducing sugar were obtained in the enzymatic hydrolysate. But these conditions also caused more degradation: per g total sugar released, 4-6 mg HMF, 16-19 mg furfural, and 160-200 mg acetic acid were found in hydrolysate. With shorter pretreatment time, the explosion effect was observed and degradation was even less: per g total sugar released, 1-2 mg HMF, 5-6 mg furfural and 65-75 mg acetic acid were formed. Results show that HMF and furfural production by this pretreatment method is very low and the acetic acid produced is below the minimum inhibitory concentration for yeast. In conclusion the supercritical CO_2 based pretreatment is very effective for preparing guayule biomass for enzymatic hydrolysis, to give high fermentable sugar yield and very low degradation to inhibitory compounds and it does not involve use of hazardous solvent.

373396 Determination of Cellulose Degree of Polymerization in Lignocellulosic Biomass

Abstract:

Determination of Cellulose Degree of Polymerization in Lignocellulosic Biomass

Lu Liu*, Hou-min Chang, Hasan Jameel, Sunkyu Park

Department of Forest Biomaterials, North Carolina State University, Raleigh, NC 27695, USA

* Presenting author: llui@ncsu.edu

Degree of polymerization (DP) is one of the most important structural properties of cellulose. Cellulose chain length expressed as cellulose DP can affect the mechanical properties of composite materials and paper products, the solubility of cellulose in a given solvent, and the efficiency of enzymatic hydrolysis of lignocellulosic biomass. Hence it is critical to establish an accurate method to determine the cellulose DP. However, current measurement technique has a major limitation that non-cellulosic components such as lignin and hemicellulose can inhibit the accurate measurement of cellulose DP. In order to overcome this limitation, we present a method with lignocellulosic biomass as a starting material using carbanilation reaction and GPC-based molecular weight (MW) distribution measurement. Delignification, alkaline extraction, and $NaBH_4$

ensure full carbanilation reaction without significant reduction in chain length. The study showed that the removal of ~ 90% of lignin by acid chlorite followed by $NaBH_4$ reduction can reduce the possibility of cellulose chain cleavage in carbanilation reaction. The alkaline extraction can remove most of the hemicellulose and the rest of lignin, which can provide an accurate result of cellulose DP. This method can be applied to both softwood and hardwood samples, which can make it a general methodology in cellulose DP determination for lignocellulosic biomass.

Abstract Submission to Be Considered As An Oral Presentation at AICHE 2014 Annual Meeting in Biomass Characterization, Pretreatment, and Fractionation Session

Quantifying Variations in the Biomass Recalcitrance of Douglas Fir

Scott Geleynse[1], Keith Jayawickrama[2], Xiao Zhang[1]
[1] Washington State University, [2] Oregon State University

To assist in the development of an economically effective industry for the production of biofuels and other products from douglas fir forestry wastes, this study discussed in this presentation aims to identify and characterize variations in the recalcitrance between individual Douglas fir trees. Characterizing these variations is intended to provide a foundation for further research into the nature of biomass recalcitrance and to determine the feasibility of potential tree improvement programs aimed at reducing recalcitrance of D. fir. In this presentation, we will discuss a method devised to screen highly recalcitrant woody biomass samples for properties related to biomass recalcitrance. The testing and application of this method using core samples taken from D. fir trees plantation grown in Oregon will also be discussed. We will present a "recalcitrance factor" that was designed to generate a single quantity to indicate the overall recalcitrance of a single tree, comparable to other trees in the same population. This factor is utilized to provide information for tree improvement programs and the screening of trees for further studies investigating biomass recalcitrance.

Presenting author:
Scott Geleynse
Bioproducts Sciences and Engineering Laboratory
School of Chemical Engineering and Bioengineering, Washington State University
2710 Crimson Way, Richland, WA 99354

Contact: scott.geleynse@email.wsu.edu, x.zhang@wsu.edu

381212 Defeating Biomass Recalcitrance with a Novel Co-Solvent Pretreatment Technology

Abstract:

Biomass recalcitrance is a major obstacle to economic conversion of lignocellulosic biomass to fuels and renewable chemicals that could sustain current and future energy demands. Recalcitrance is a multi-layered phenomenon influenced by physical and chemical characteristics of biomass that prevent enzymes from accessing and hydrolyzing its sugar polymers into fermentable monomers. We recently developed a highly-effective pretreatment strategy that uses tetrahydrofuran (THF) as a renewable co-solvent to enhance both biomass deconstruction and delignification and achieve unprecedented sugar yields from raw agricultural and forestry feedstocks. By optimizing THF co-solvent pretreatment of corn stover, we achieved nearly theoretical yields of glucose with as little as about 4 FPU of enzyme/gram of glucan in the corn stover. We report here physiochemical changes in biomass that result from THF pretreatment to understand factors that could be responsible for its unprecedented digestibility. We compared results from THF co-solvent pretreatment for three different biomass feedstocks (corn stover, maple wood, and poplar wood) to those from application of leading ethanol-organosolv and dilute-acid options. The pretreated solids were first characterized by SEM and NMR to follow differences in micro- and macro- level structures and crystallinity. Simons' Staining as well as long-term enzyme activity studies were also employed to better understand substrate-enzyme interactions. These results showed that THF co-solvent was highly effective in solubilizing both hemicellulose and lignin fractions to produce solids highly enriched in glucan that were more amorphous than from other pretreatments. In addition to being highly accessible to enzymes, less enzyme deactivation was observed over long hydrolysis times, suggesting the loss of enzyme activity was not a major factor limiting yields from biomass. Overall, these findings further support the potential value of THF co-solvent pretreatment as a promising strategy for overcoming biomass recalcitrance.

Pretreatments as Tools to Improve Fast Pyrolysis Products from Lignocellulosic Biomass

René Garrido, Joseph Reckamp, Nicole Hammer, Charles Coe, Justinus A. Satrio*

Department of Chemical Engineering, Villanova University, Villanova, PA 19085

Extensive resources and efforts have been dedicated to assess opportunities for using locally available biomass to produce energy, reduce CO_2 emissions, and lower reliance on petroleum.. Fast pyrolysis is one promising approach for converting lignocellulosic biomass to produce chemicals and fuels that has gained interest in the past few years. Although fast pyrolysis has been known and studied for many years; due the complexity of fast pyrolysis reaction mechanisms there are still a lot more to investigate to increase the understanding on fast pyrolysis of biomass. The increased interest in utilizing bio-oil to produce renewable transportation fuels, knowledge about its chemical compositions and how it could be controlled is crucial for developing cost-effective bio-oil based technologies.

The core objective of this research is to study the influence of pretreatments (acid hydrolysis, torrefaction, and inorganic salts impregnation) on selected lignocellulosic biomass to improve fast pyrolysis products. It was found that inorganic matters inherent in biomass affect the selectivity towards certain chemical components contained in the bio-oil. Acid hydrolysis and torrefaction were investigated for their ability to reduce the metal-induced formation of organic acids, ketones and aldehydes, which destabilizes the bio-oil of untreated biomass. Addition of phosphoric acid changes the reaction pathways from levoglucosan (LG) to levoglucosenone (LGO) in treated *Phragmites australis* (PHA) and Paper Mill sludge (PMS). On the other hand, biomass with low inorganic content, such as pinewood, can also be treated to increase the quality of the bio-oil by addition of a Lewis acid, such as $ZnCl_2$. This Lewis acid promotes the production of furfural (FF) and LGO while reducing the formation of organic acids, furans and ketones and results in a Zn-rich char that aids its conversion to a higher value activated carbon.

*Contact person: justinus.satrio@villanova.edu

390570 Cellulose Reducing Ends Have No Impact on Cellulose Biological Conversion with T. Reesei Cellulase

Abstract:

Trichoderma reesei derived cellulase cocktail is comprised of several cellobiohydrolases (Cel 7A or CBH I; Cel6A or CBH II) and endoglucanases (EG I through V). Earlier studies showed that the two cellobiohydrolases CBH I and CBH II act preferably from the reducing and non-reducing ends of cellulose, respectively. Thus, it has long been believed that cellulose reducing ends (REs) play a vital role in cellulose enzymatic hydrolysis mediated by fungal enzymes [1,2]. However, the fundamental hydrolysis studies investigating the effect of REs were mostly conducted with soluble cellodextrins, and studies applying insoluble substrate are limited. In one such report, Xu et al. recently reported that oxidation of REs and primary hydroxyl groups (PHGs) significantly affected cellulose conversion [3]. However, the quantification of cellulose REs following oxidation were not reported in this study. On the contrary, a few other studies reported a limited effect of REs on cellulose conversion [4]. Therefore, in this study, the role of REs in insoluble cellulose recalcitrance to biological conversion with *T.ressei* derived enzymes was investigated. Pure cellulose compounds were subjected to sodium borohydride ($NaBH_4$) reduction and sodium chlorite-acetic acid (SC/AA) and bicinchoninic acid (BCA) oxidation to block the REs. $NaBH_4$and SC/AA treatments were able to block more than 95% and 60% of the REs, respectively, than untreated controls, while BCA applied for various cycles oxidized cellulose to various extents. X-ray diffractograms of the treated solids showed that these treatments did not change the cellulose crystalline structure, and crystallinity index determined by Segal's and amorphous subtraction methods was unchanged. Enzymatic hydrolysis of resulting solids with complete cellulase system at protein loadings as low as 0.5 mg/g glucan revealed that cellulose REs did not affect initial or final cellulose conversion. Furthermore, in contrast to what is traditionally believed, blocking cellulose REs had no impact on cellulose conversion with purified cellulase components (CBH I, CBH II, and EG I) and did not affect their synergy. On the other hand, oxidation of the PHGs significantly affected cellulose digestibility, with the adverse effect was more prominent at low cellulase loadings.

1.Teeri, T. T. *et al.* Modes of action of two Trichoderma reesei cellobiohydrolases. *Prog. Biotechnol.* **10**, 211-224 (1995).

2. Zhang, Y. H. P. & Lynd, L. R. A functionally based model for hydrolysis of cellulose by fungal cellulase. *Biotechnol. Bioeng.* **94**, 888-898 (2006).

3.Xu, F., Ding, H. & Tejirian, A. Detrimental effect of cellulose oxidation on cellulose hydrolysis by cellulase. *Enzyme Microb. Technol.* **45**, 203-209 (2009).

4.Kurašin, M. & Väljamäe, P. Processivity of Cellobiohydrolases Is Limited by the Substrate. *J. Biol. Chem.* **286**, 169-177 (2011).

Abstract Submission to Be Considered as an Oral Presentation at ***AICHE 2014 Annual Meeting*** in *T4A06 Biochemical Conversion Processes in Forest/Plant Biomass Biorefineries II section*

Chemical Conversion of Biorefinery Lignin to Value Added Platform Chemicals.

Ruoshui Ma[1], Vince Hebert[1], Jinwen Zhang[1], Tom Candy[2], Wei Wang[2], Xiao Zhang[1]

1. Washington State University, Richland, WA, 99354
2. Solvay Chemicals Inc., Houston, TX, 77098

Lignin is the largest nature source of renewable material with an aromatic skeleton. Depolymerizing lignin to low molecular weight aromatic and phenolic compounds offers attractive opportunities to produce a range of high value products/chemicals. Lignin can be depolymerized into low molecular weight phenolic compounds (LMWPC) by a number of chemical and/or biological means. Among these pathways, oxidation strategy has shown its advantages to selectively increase functionality of the final products.

In this presentation, we will report recently work from our group on applying peroxygen chemicals to depolymerize and convert biorefinery lignin into platform low molecular weight phenolic and quinone compounds.

Presenting author: Ruoshui Ma, Chemical Engineering, Washington State University, Richland, WA, 99354

Contact: ruoshui.ma@email.wsu.edu

xiaozhang@tricity.wsu.edu

Abstract Submission to Be Considered as an Oral Presentation at AICHE 2014 Annual Meeting in T4A03 Chemical Conversion Processes in Forest/Plant Biorefineries I session

Metal catalyzed conversion of Biomass to Maximize Hemicellulose Sugar Conversion to Valuable Organic Acids

Carlos Alvarez-Vasco, Xiao Zhang

Hemicellulose usually represents between 22 to 30% of biomass dry weight been along with cellulose and lignin the main biomass constituents. However, due to its lower degree of polymerization and crystallinity compared to cellulose, hemicellulose is more sensitive to thermal and chemical degradation during biomass pretreatment. With current pretreatment technologies, hemicellulose degradation compounds do not add value to the overall process. In fact, they may even be detrimental for subsequent fermentation or purification steps.

This study investigated the potential of metallic ions addition during pretreatment in order to maximize hemicellulose conversion to valuable chemicals and cellulose recovery to fermentable sugars. Main benefices of metallic ions addition include its potential higher selectivity toward specific products and the possibility of been recovery and use in several runs. In addition, the natural occurrence of metallic ions in biomass structure (K and Ca on ash content) or in forest residues (Al and Fe due to soil contamination) are a potential source of this ions.

Biomass metallic ion catalysis takes advantage of the natural occurrence of these metallic ions to catalyze the conversion of Hemicellulose to organic acids. Major conversion pathways involved in carbohydrate metal catalyzed conversion of biomass and hemicellulose model compounds (i.e. glucomannan and xylan) are elucidated. Considerable amounts of carboxylic acids including lactic acid, glycolic acid, succinic acid and formic acid are produced along with pretreated biomass. The addition of metallic ions during pretreatment increased the formation of products for both biomass and tested carbohydrate polymers model compounds. In addition, the role of metallic ion catalyzed pretreatment conditions on resulting enzymatic cellulose-to-glucose yields of pretreated biomass are study. This study demonstrated an integrated metallic ion treatment of biomass with the opportunity to produce valuable chemicals from hemicellulose in order to maximize the value of biomass carbohydrates.

Presenting author: Carlos Alvarez-Vasco

Bioproducts Science Engineering Laboratory, Washington State University, 2710 Crimson Way, Voiland School of Chemical Engineering and Bioengineering, Richland, WA 99354

Contact: c.alvarezvasco@email.wsu.edu, xiaozhang@tricity.wsu.edu

Guangzhou Institute of Energy Conversion， Chinese Academy of Sciences

Catalytic Hydroprocessing of Fatty Acid Methyl Esters to Jet Fuels

Pengmei Lv

Marriott Marquis Atlanta,

November 17, 2014,

Chemical Conversion Processes in Forest/Plant Biorefineries

AIChE The Global Home of Chemical Engineers

Table of contents

1. **Background**
2. **Molecular sieve synthesis and characterization**
3. **Catalytic performance**
4. **Conclusion**

1. Background

Problems related to biodiesel

- ☐ A higher cold filter plugging point;
- ☐ Containing oxygen, less stable;
- ☐ Lower heating value

↓

Catalytic hydrotreated and isomerized to bio-aviation fuel

State of the art of the catalysts

➢**Microporous Si-Al molecular sieve:** Relative strong acidity and good hydrothermal stability, whereas its micro pore size impedes the diffusional mass transfer of reactants and products.

➢**This case:** The molecular size of FAME reaches over 4.4 nm.

↓

Demand: **A kind of molecular sieve with larger pore size, namely mesoporous molecular sieve is in need.**

2.Molecular sieve synthesis and characterization

(1) Introduction

(2) Synthesis

Principle, catalyst design, catalyst preparation

(3) Characterization

SEM, XRD, NMR, surface area, pore volume, diameter and acidity

(1) INTRODUCTION

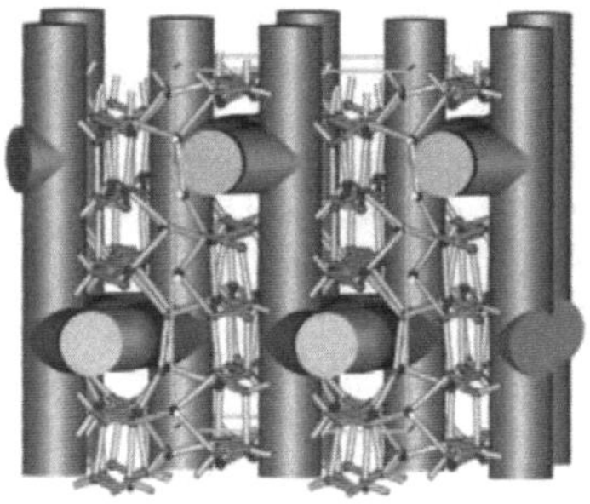

- **Definition:** As defined by IUPAC (International Union of Pure and Applied Chemistry), mesoporous materials means their pore size ranges between 2-50 nm.
- **Molecular sieve:** A class of materials with uniform and ordered porous structure .
- **Typical spatial structure:** Cage-like ordered arrangement.
- **The first ordered mesoporous SiO_2 material :** Synthesized by Mobil company in 1992.
- **Generally, two categories:** Silicon-based mesoporous material and non-silicon based mesoporous material.

(2) SYNTHESIS

➢ Synthesis principle

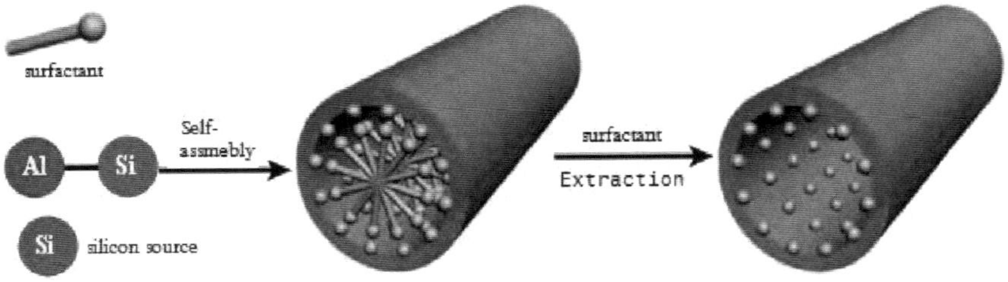

Supramolecular assembly through cooperative effect of hybrid organic - inorganic species.

➢ Catalyst design

(1) **Carrier** – Mesoporous molecular sieve, dependent upon the size scale of FAME molecuar. Surfactant selection and hydrothermal crystallization process investigation。

(2) Main active metal component and complex component- Ni, the main active component for hydrotreating reactions, M selected to be transition metal, e.g. Co, Zr, etc.

(3) **Appropriate Bronsted and Lewis acid sites distribution** – Involved reactions include FAME hydrotreatment, isomerization and crack, etc, achieved through the Si/Al ratio regulation and preparation process.

(3) CHARACTERIZATION

➢ SEM characterization

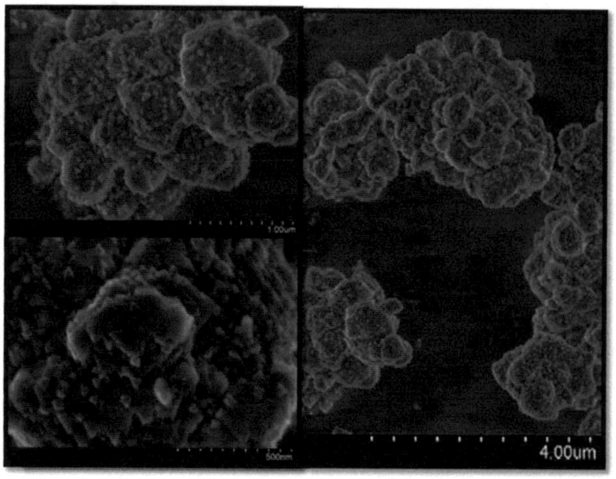

A serrated surface of 10wt%Ni/Hβ

➢ XRD characterization

2θ/(°)	Interplanar spacing, D/nm	relative intensity
7.8	1.14±0.02	MS
12.0	0.74±0.02	W
13.6	0.67±0.02	W
21.0	0.425±0.01	W
22.4	0.397±0.01	VS
29.4	0.30±0.01	W
41.0	0.22±0.01	W

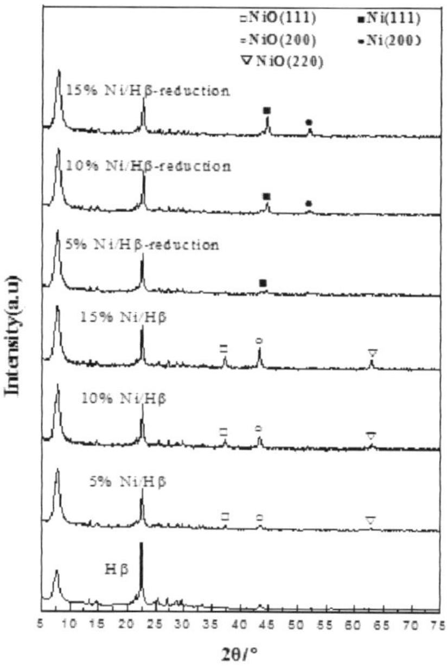

➢ After Ni loaded, although the intensity changes, its crystal structure was maintained well;

➢ With the increase of Ni loading, NiO diffraction peak becomes more and more strong.

➢ NMR characterization

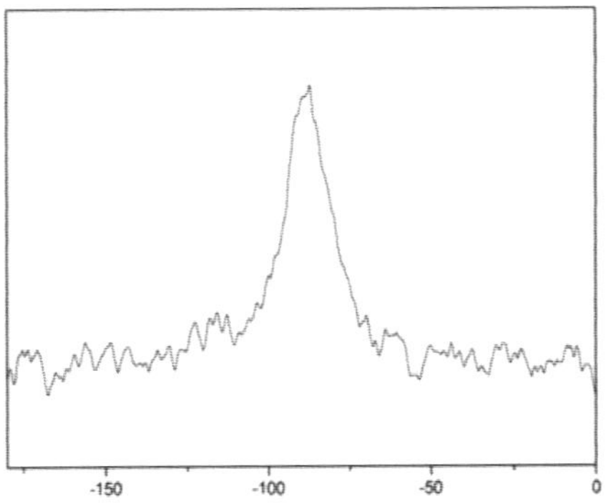

Al/SBA-15 ^{29}Si MAS NMR

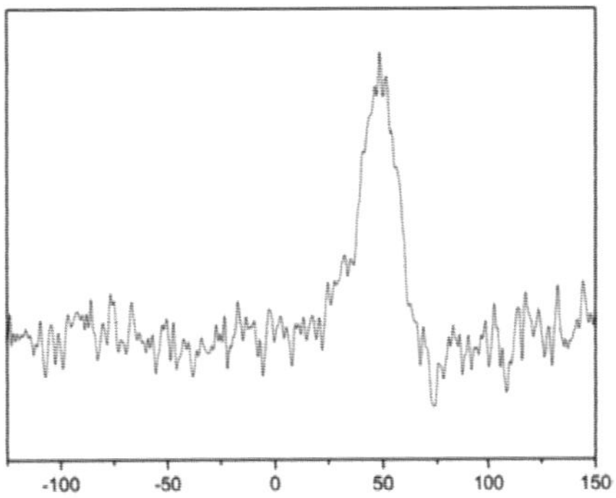

Al/SBA-15 ^{27}Al MAS NMR

➢ NMR shows that Al appears a four-coordinated form in the framework of molecular sieve to generate Lewis acid sites.

➢ Surface area, pore volume, diameter and acidity

Catalyst	Ni loading (wt%)	Si/Al	Physical structures: BET surface area ($m^2 \cdot g^{-1}$)	Pore volume ($cm^3 \cdot g^{-1}$)	Pore diameter (nm) (HK method)	Acid density ($mmol \cdot g^{-1}$): Weak acid sites	Strong acid sites	Total amount
Hβ	0	25	554.4	0.20	0.66	0.31	0.30	0.61
Ni/Hβ	5	25	511.7	0.19	0.63	0.39	0.33	0.72
Ni/Hβ	10	25	483.4	0.18	0.62	1.05	0.34	1.39
Ni/Hβ	15	25	458.1	0.17	0.60	0.85	0.18	1.03
Ni/Hβ	10	50	462.2	0.17	0.62	0.92	0.22	1.14
Ni/Hβ	10	150	445.8	0.17	0.62	0.83	0.13	0.96
SiO_2	0	/	356.5	0.84	9.6(BJH method)	0	0	0
Ni/ SiO_2	10	/	315.9	0.72	8.5(BJH method)	0.11	0.09	0.20

- NiO particles located on the surface of Hβ create a weakly acidic environment after H_2 reduction forming the H^+ as Brønsted acid ($Ni^{2+} + H_2 = Ni + 2H^+$).
- On the other hand, the particles of Ni may block the Hβ microchannels, thus decreasing the accessibility of NH_3.

3. Catalytic performance

(1) Different Ni loading and Si/Al ratio

(2) Different temperatures

(3) Different time and reusability test

The composition of FAMEs

FAMEs	Fatty acid compositions[a] of FAMEs structure	Formula	Compositions (wt%)
Methyl Myristic	$C_{14:0}$[b]	$C_{15}H_{30}O_2$	0.35
Methyl Hexadecanoate	$C_{16:0}$	$C_{17}H_{34}O_2$	21.22
Methyl Octadecanoate	$C_{18:0}$	$C_{19}H_{38}O_2$	33.74
Methyl Oleate	$C_{18:1}$	$C_{19}H_{36}O_2$	39.52
Methyl Linoleate	$C_{18:2}$	$C_{19}H_{34}O_2$	0.19
Methyl Linolenate	$C_{18:3}$	$C_{19}H_{32}O_2$	4.23
Methyl Arachidate	$C_{20:0}$	$C_{21}H_{42}O_2$	0.75

[a] Fatty acid composition was assigned on the basis of alkyl chain length.
[b] Nomenclature refers to the number of carbon atoms and number of C=C double bonds.

Hydrotreatment of FAMEs was carried out in a 100 mL stainless slurry reactor equipped with a polytetrafluoroethylene (PTFE) liner and mechanical stirrer operating in batch mode.

(1) Different Ni loading and Si/Al ratio

Catalyst	Ni loading (wt%)	Si/Al	Conversion (%)	Selectivity (%)						
				C_{15}	iso-C_{16}	C_{16}	C_{17}	iso-C_{18}	C_{18}	C_{15}–C_{18}
Hβ	0	25	3.2	1.2	0	23.9	5.5	0	65.8	96.4
Ni/Hβ	5	25	95.3	4.0	0.8	14.1	14.6	2.7	52.7	88.9
Ni/Hβ	10	25	100	4.7	1.2	14.1	17.1	3.6	52.5	93.2
Ni/Hβ	15	25	98.7	7.6	0.9	9.2	27.2	2.6	30.3	77.8
Ni/Hβ	10	50	97.2	2.2	0.5	17.5	10.1	1.8	61.3	93.4
Ni/Hβ	10	150	96.8	0.7	0.1	19.0	7.6	0.5	66.1	94.0
Ni	Pure Ni	/	37.2	26.9	0	0.7	66.3	0	0.2	94.1
Ni/SiO_2	10	/	91.2	19.8	0	1.6	70.6	0	2.7	94.7

Reaction conditions: FAMEs (1.0 g); dodecane (4.0 g); catalyst (0.1 g); 270 °C; 1 MPa H_2; 8 h.

- A pure Hβ support without Ni loading , very low conversion of 3.2%.
- Comparatively, for pure Ni, a higher conversion of 37.2%. This proves that Ni alone can catalyze hydrotreatment of FAMEs to alkanes without support, including hydrogenation, hydrogenolysis, hydrodecarboxylation reactions, et al.
- Hβ can provide a larger catalytic specific surface area and more contact sites to the reactant than pure Ni metal.
- For Ni/Hβ, compared with other products, the content of C15 and C17 is relative low. This proves that the catalytic hydrotreatment occurs mainly via HDO without carbon loss.

Catalyst	Ni loading (wt%)	Si/Al	Conversion (%)	Selectivity (%)						
				C_{15}	iso-C_{16}	C_{16}	C_{17}	iso-C_{18}	C_{18}	C_{15}–C_{18}
Hβ	0	25	3.2	1.2	0	23.9	5.5	0	65.8	96.4
Ni/Hβ	5	25	95.3	4.0	0.8	14.1	14.6	2.7	52.7	88.9
Ni/Hβ	10	25	100	4.7	1.2	14.1	17.1	3.6	52.5	93.2
Ni/Hβ	15	25	98.7	7.6	0.9	9.2	27.2	2.6	30.3	77.8
Ni/Hβ	10	50	97.2	2.2	0.5	17.5	10.1	1.8	61.3	93.4
Ni/Hβ	10	150	96.8	0.7	0.1	19.0	7.6	0.5	66.1	94.0
Ni	Pure Ni	/	37.2	26.9	0	0.7	66.3	0	0.2	94.1
Ni/SiO_2	10	/	91.2	19.8	0	1.6	70.6	0	2.7	94.7

For pure Ni or supported on SiO_2 ,

- A relative higher content C_{15} and C_{17} is obtained. This indicates the catalyst with no or little acidity may favor hydrodecarboxylation or hydrodecarbonylation reaction.
- The iso-alkanes content is 0. On the account of no or little acidity of these 2 catalysts, this proves that the acidic sites are the active centers for isomerization reaction.

(2) Different temperatures

Temperature (°C)	Conversion (%)	Selectivity (%)						
		C_{15}	iso-C_{16}	C_{16}	C_{17}	iso-C_{18}	C_{18}	C_{15}–C_{18}
230	49.4	0.8	0	18.3	7.2	0	72.4	98.7
240	62.4	1.3	0.4	17.4	8.3	0.7	70.6	98.7
250	70.5	1.7	0.7	16	10.5	1.6	68	98.5
260	97.4	2.9	1	15.6	13.1	3	60.5	96.1
270	100	4.7	1.2	14.1	17.1	3.6	52.5	93.2
280	100	4	0.8	12.8	13.6	3.2	43.7	78.1

Reaction conditions: FAMEs (1.0 g); dodecane (4.0 g); catalysts (0.1 g); 1 MPa H_2; 8 h

- A higher reaction temperature promotes not only the hydrotreatment of FAMEs – especially the hydrodecarbonylation or hydrodecarboxylation pathway – but also the enhanced cracking of long-chain alkane products.

(3) Different time and reusability test

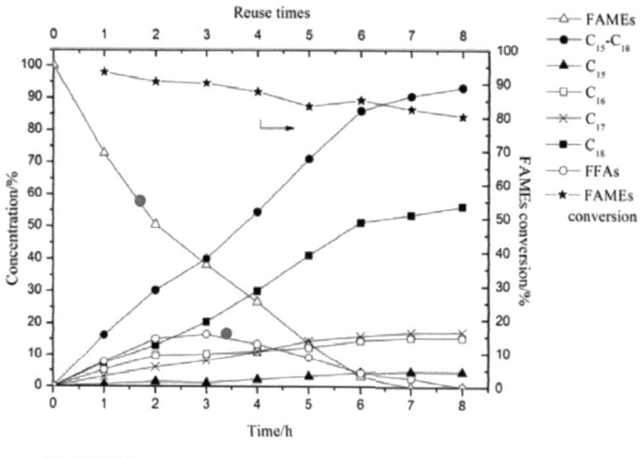

FAME hydrotreatment catalyzed by 10%Ni/Hβ

- The conversion of FAMEs in 1H decreased 27.2%. This illustrates the hydrogenation reduction of unsaturated bond is a rapid process.
- The intermediate product of FFAs reached the highest concentration of 16.5% at the 3rd hour. In the first 3H, FFAs accumulation rate was higher than that of its conversion rate. After 3h, FFAs was converted into alkanes.

4. Conclusion

Hydrotreatment and isomerization pathway

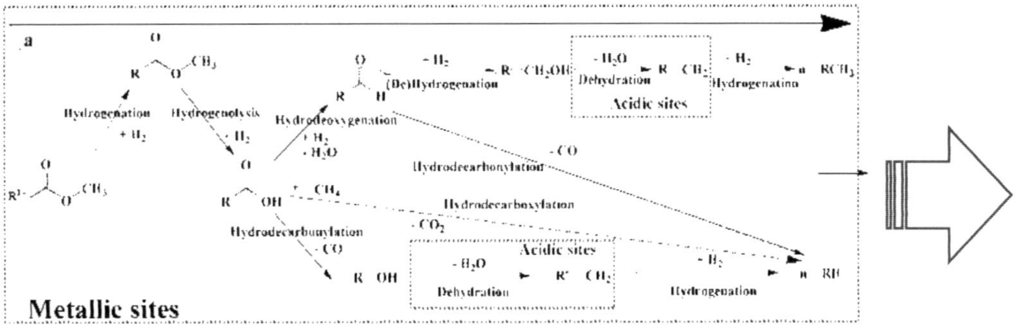

(1) Metal active center catalyzed hydrogenation of unsaturated bonds in the alkyl chain.

(2) Hydrogenolysis of the formed saturated FAMEs to produce carboxylic acids, e.g., hexadecanoic and octadecanoic acids.

a. subsequent hydrodeoxygenation of carboxylic acids to aldehydes then to alcohols, followed by dehydration via acidic functionalities in the zeolite catalyst to alkenes and subsequent hydrogenation to alkanes.

b. carboxylic acids directly change to alkanes via decarboxylation with carbon loss.

c. carboxylic acids undergo decarbonylation to alkanes.

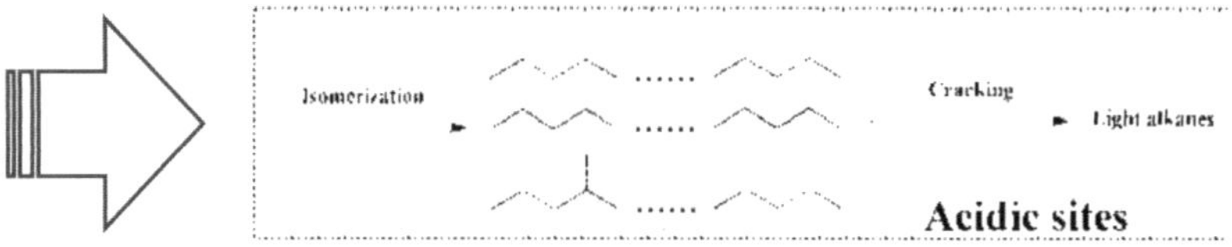

(3) All long-chain alkanes could be subjected to hydroisomerization and hydrocracking at the function of acidic sites on the catalyst.

(4) Carbon dioxide and carbon monoxide may react with H_2 to produce methane and water.

① The amount of acid sites increase with Al content. The acid sites promote the isomerization reaction.

② The catalyst with no or little acidity may favor hydrodecarboxylation or hydrodecarbonylation reaction.

③ Metallic sites are the active center for hydrodeoxygenation. A higher reaction temperature promotes the cracking of long-chain alkane products.

④ With certain range of metal loading and acid amounts, the catalyst will show an optimal catalytic performance on conversion and selectivity.

⑤ 10 wt% Ni/Hβ with Si/Al = 25 shows the best performance both for the deoxygenation reaction and isomerization reaction.

Hydrotreated alkanes

Palm oil biodiesel

Micro pore ZSM-5

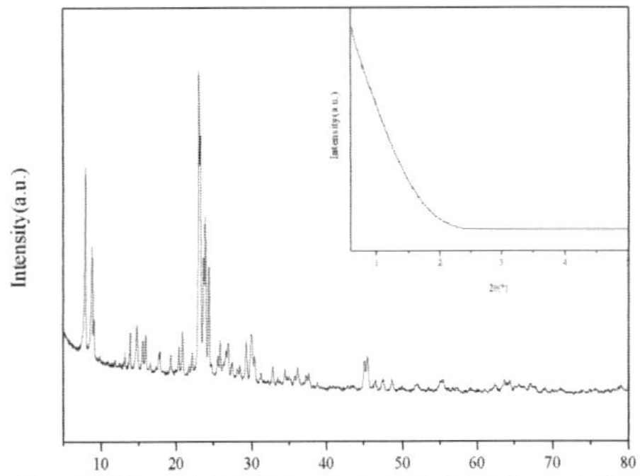

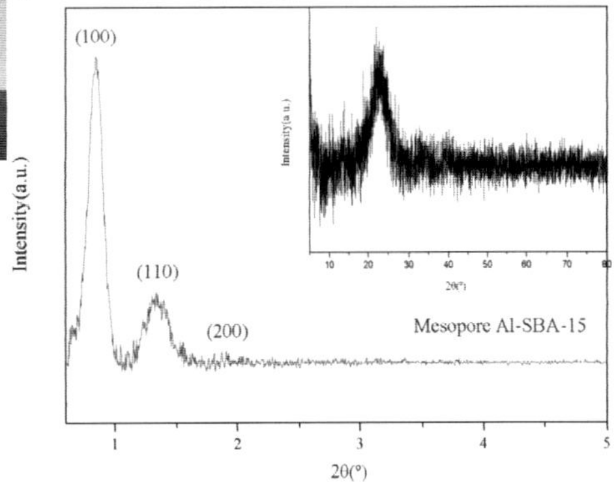

Mesopore AL-SBA-15

Thanks for your attention !

389277 Saccharification and Lignin Fractionation of Lignocellulosic Biomass in Molten Salt Hydrate System

Abstract:

Molten salt hydrate (MSH) has unique properties, such as high boiling point and depressed freezing point, very low water vapor pressure, enhanced acidity, and high enthalpy of mixing. Some MSHs are able to swell and solubilize cellulose through breaking inter- and intra-molecule hydrogen bonds of cellulose. Compared to ionic liquids, MSHs are inexpensive and easier to prepare and recover. Recently, our investigation indicated that lignocellulosic biomass (such as corn stover, switchgrass, hardwood and softwood) could be quickly and completely saccharified without any pretreatment under moderate conditions (110-130 °C for 10-60 min) in molten salt hydrate systems. Both cellulose and hemicellulose in lignocellulosic biomass were completely hydrolyzed into fermentable sugars with high yield and selectivity (>90%). Lignin in the biomass was remained as insoluble residue and could be easily separated for co-products development. In this presentation, we report the results of process optimization of biomass saccharification in molten salt hydrate, sugar-salt separation after the saccharification, lignin characterization and co-product potential, and techno-economic analysis of the system.

372234 Conversion of Lignin to Hydrocarbon Fuels Via Catalytic Depolymerization and Hydrogenation

Abstract:

Three different solvent systems (sodium hydroxide solution, ethanol aqueous solution and aqueous solution of phosphoric acid) were chosen to study the conversion of lignin to hydrocarbons via catalytic depolymerization and hydrogenation. A two-step method, selective depolymerization of lignin to liquid compounds rich in phenolics followed by hydrogenation of the obtained liquid, was developed. It was found that sodium hydroxide solution showed the best performance among the three, as the yield of phenolic compounds could reach 60%, with a selectivity of 85% when used for lignin degradation. What's more, the selectivity for hydrocarbons could reach 97% when the derived phenolic compounds were hydrogenated. The preliminary mechanism study using model compounds indicated that the conversion of lignin to hydrocarbons is a multi-step process consisting of alkaline hydrolysis, demethylation, demethoxylation and transalkylation.

373657 Catalytic Conversion of Xylose-Rich Hydrolyzate from Autohydrolysis into Furfural

Abstract:

Catalytic conversion of xylose-rich hydrolyzate from autohydrolysis into furfural

Lu Liu*, Jing Du, Hou-min Chang, Hasan Jameel, Sunkyu Park

Department of Forest Biomaterials, North Carolina State University, Raleigh, NC 27695, USA

* Presenting author: llui@ncsu.edu

Furfural, the dehydration product of xylose, can be served as a sustainable substitute for petroleum-based building blocks used in production of fine chemicals and plastics. Aiming at the improvement of current method to produce furfural from xylose, our study is focused on an efficient and environmentally friendly method to produce furfural using the autolydrolysis hydrolyzate from lignocellulosic biomass. The optimization process of temperature, pressure, and other experimental conditions in lab-scale are described based on xylose conversion and furfural yield in both systems, pure xylose solution and biomass hydrolyzate. Different types of solid catalyst and solvent system were applied to compare with non-catalytic scenario to achieve high yield of furfural production. In model system of pure xylose solution, the furfural yield was not significantly improved by adding catalyst, while the xylose conversion yield was significantly increased. About 40% furfural yield was achieved in 2% xylose solution after 3 hours of reaction in non-catalytic scenario with about 30% xylose left. With biomass hydrolyzate from autohydrolysis, an enhanced furfural yield and xylose conversion yield was achieved. To maximize conversion yield of xylose, a reactor was modified to allow the produced furfural and steam to be released and collected. The challenges and opportunities of hydrolyzate conversion into furfural will be discussed in details, considering hydrolyzate impurities, conversion mechanism, and scale-up feasibility.

387447 Unveiling the Ionic Liquid Deconstruction of Lignocellulosic Biomass Using Glycome Profiling

Abstract:

Certain ionic liquids (ILs) are capable of effectively overcoming plant cell wall recalcitrance; however, the underlying mechanisms are not fully understood. To better understand how certain ionic liquids deconstruct biomass, we applied an immunological approach ("glycome profiling") that employs a comprehensive suite of plant glycan-directed monoclonal antibodies to monitor cell wall structural/compositional/extractability changes in switchgrass biomass. A range of ILs with very diverse chemical and solvent properties was tested under typical pretreatment conditions, with 1-butyl-3-methylimidazolium chloride ([C_4mim][Cl]) solubilizing and regenerating cellulose, N-ethyl-N-(furanyl-2-methyl) ethanamine dihydrogen phosphate ([FurEt$_2$NH][H_2PO_4]), N-ethyl-N-(4-methoxybenzyl) ethanamine dihydrogen phosphate ([p-AnisEt$_2$NH][H_2PO_4]) and cholinium lysinate ([Ch][Lys]) extracting large fraction of lignin, and 1-ethyl-3-methylimidazolium acetate ([C_2mim][OAc]) solubilizing both lignin and cellulose. Glycome profiling studies on pretreated biomass residues, in combination with wet chemistry quantification and ^{13}C CP/MAS NMR and 2D HSQC NMR analysis, revealed distinct changes in the composition and extractability of non-cellulosic cell wall glycans, suggesting different deconstruction mechanisms caused by different ILs. Interestingly, all ILs induced the disruption of lignin-polysaccharide interactions as reflected by a loss of pectins and arabinogalactan epitopes in less harsh extracts (oxalate and carbonate) and significantly increased extractability of hemicellulosic glycans in oxalate, carbonate, 1M KOH and 4M KOH extracts. This study establishes a deeper scientific understanding of IL pretreatment and provides insights on the rational design of task-specific ILs for biomass deconstruction.

366787 Recycling of Inexpensive Ionic Liquids for Lignocellulose Fractionation

Abstract:

Lignocellulosic biomass is a composite material comprising cellulose, hemicellulose and lignin and is a target feedstock for the bio-based chemicals and material economy. We have developed low-cost ionic liquids that are alternatives to aqueous or organic solvent deconstruction (aka pretreatment) systems, with distinct processing advantages in the separations, such as excellent separation of cellulose from lignin and recovery of a valuable lignin fraction.

Previous IL-based biomass research has focused on swelling and dissolving the entire lignocellulose with dry ILs. The requirement for costly ionic liquids has resulted in challenges for IL recovery. Water, a major component of biomass, prevents cellulose dissolution, necessitating extensive drying of biomass and the hygroscopic IL, increasing energy costs.

We have overcome this by redesigning the process for lignin dissolution, yielding filterable cellulose, preconditioned for further use, and an IL-lignin solution for precipitation or conversion to high-value chemicals. We use a range of 'protic' ILs, the family typically used in IL industrial processes because their simple acid-base chemistry makes them easy and cheap to prepare. Despite the low cost (as low as $1.24 per kg)[1] and low environmental impact, solvent recycling is mandatory for this application.[2]

The results from this study demonstrate that the ILs maintain solvent stability under long-term processing conditions, that they can be recovered and continue to exhibit very good cellulose lignin fractionation after multiple reuses.

[1] L.Chen, M. Sharifzadeh, N. Mac Dowell, T.Welton, N.Shah, J.P. Hallett, Green Chemistry, 2014.

[2] D. Klein-Marcuschamer, B. A. Simmons and H. W. Blanch, Biofuels, Bioprod. Biorefin., 2011, 5, 562–56.

387051 Flocculation Assisted Clarification of Enzymatically Hydrolyzed Corn Stover Slurries

Abstract:

The biochemical conversion of lignocellulosic biomass to ethanol is able utilize robust microorganisms that tolerate residual solids and moderate concentrations of soluble inhibitory species. However, many potential hydrolysis-based conversion pathways to hydrocarbon fuels are highly sensitive to the composition of the hydrolyzate slurries. Removal of the insoluble solids from the slurry is a requisite first step before additional aqueous species fractionation or concentration steps may occur.

Here we present the development of a solid-liquid separation process for pretreated and enzymatically hydrolyzed corn stover slurries. The separation of the insoluble lignin residues from these slurries is challenging due to the high concentration of solids (5-10 wt%) and the fact that the solids are relatively small (23 μm mean diameter) and deformable. The focus of the work is vacuum-driven bulk filtration aided by flocculation. Without flocculation, the solids form a compressible cake with low permeability. Conversely, adding low doses of a cationic polyacrylamide flocculant generated large flocs (1 mm target size) which formed highly permeable cakes under low applied differential pressures. In general, flocculation greatly increased the overall capacity of vacuum filtration for the slurries tested, but this comes at the expense of increased wash water, where washing is used to remove residual sugars from the cake bed. Finally, we show that a polishing filtration step may be avoided when using flocculants, depending on the requirements of the subsequent upgrading process.

384608 A Comparative Study on the Pyrolysis and Gasification Behavior of Sugarcane Tops/Leaves and Sugarcane Bagasse

Abstract:

After extracting sugar at sugarcane mill, bagasse is left as a residual byproduct. Some of this bagasse is used to meet the energy requirements of the mill. The remaining bagasse could be pyrolyzed and gasified to provide additional feedstock for producing renewable fuels and chemicals. Also, a portion of the sugarcane tops and leaves ("SCT") left in the fields can also be delivered to the bagasse processing unit to provide additional feedstock for making fuels and chemicals. Often SCT is burned in the fields because there is too much SCT remaining to properly maintain the fields. Burning of SCT in the fields creates environmental problems.

The chemical composition of SCT is not the same as bagasse. This study was done to characterize differences in pyrolysis and gasification of these two feedstocks. TGA weight loss data during pyrolysis indicates that cellulose peak reactivity in SCT is shifted to a lower temperature compared to bagasse. Based on simultaneous screening by TGA and DSC, the SCT showed a lower activation energy and higher gasification reactivity in CO_2 than bagasse. This trend is consistent with the higher inorganic content in SCT since some of the inorganics present may catalyze these reactions. The heat effects associated with pyrolysis and gasification are different for these two feedstocks. Slow pyrolysis in nitrogen (at 15 K/min) indicated mildly endothermic reaction with sugarcane bagasse while the reaction is nearly heat neutral with SCT. Higher reactivity of SCT during oxidative pyrolysis (with 5% O_2 in N_2) resulted in a much higher exotherm than bagasse. The differences observed are not likely to require that the bagasse and SCT need to be pyrolyzed and gasified separately.

380688 Comparison of Lab, Pilot, and Commercial Scale Low Consistency Mechanical Refining for Improvements in Enzymatic Digestibility of Pretreated Hardwood

Abstract:

Mechanical refining has been shown to improve biomass enzymatic digestibility. In this study industrial high-yield sodium carbonate hardwood pulp and was subjected to lab, pilot and industrial refining to determine if the mechanical refining improves the enzymatic hydrolysis sugar conversion efficiency differently at different refining scales. Lab, pilot and industrial refining increased the biomass digestibility for lignocellulosic biomass relative to the unrefined material. The sugar conversion was increased from 47% to 72% at 5 FPU/g of biomass with refining at 67.0 kWh/t. There is a maximum in the sugar conversion with respect to the amount of refining energy. Water retention value is a good predictor of improvements in sugar conversion for a given fiber source and composition. Improvements in biomass digestibility with refining due to lab, pilot plant and industrial refining were similar with respect to water retention value. The effects of various mechanical refining operating conditions will be discussed in detail. These conditions including refiner plate design, refiner plate gap, refining energy input and other energy considerations, refining consistency (especially high consistency refining), and refiner operation costs and economics. Fundamental studies of how particle size and surface area change with refining and how different fractions contribute toward the enzymatic hydrolysis efficiency will also be presented.

Purification and concentration of wood extracts by membrane filtration

By Yipeng Xie
Advisor: Dr.Shijie Liu
Department of Paper and Bioprocess Engineering
SUNY college of ESF

Outline

1.Background
2.Equipment and Procedures
3.Mathematical model
4.Results analysis and discussion
5.Conclusion
6.Reference

1. Background

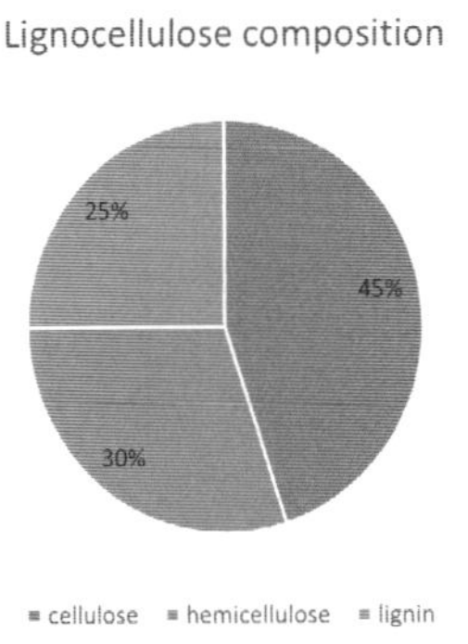

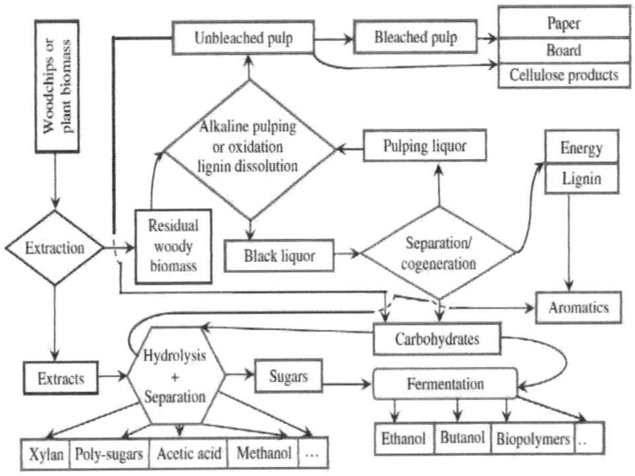

Composition of hot water extracts

Hot water extraction of *Paulownia* woodchips was performed at 160 °C for 2 h, with a solid to liquor ratio of 1:8.

Composition: lignin, polysaccharides, furfural, acetic acid, etc.

Fermentation-Inhibitors	Methanol	Hydroxyl methyl-furfural(HMF)	Formic acid	furfural	free acetic acid, mol/L	bound acetic acid, mol/L
Concentration, g/L	0.8339	0.4613	0.3092	0.5186	0.03189	0.004113
sugars	galactose	glucose	xylose	arabinose	rhamnose	mannose
Concentration, g/L	1.603882	1.381168	1.513271	0.796788	trace	trace

Separation methods

Separation Method	Equipment cost	Operation cost	Recover cost for valuable products	Efficiency	Environmental harm
Distillation	High	High	Low	High	Low
Chemical extraction	Low	Medium	High	Medium	High
Filtration	Medium	Medium	Low	High	Low
Centrifugation	High	High	Low	Low	Low
Vacuum evaporation	High	High	Low	Medium	Low
Chromatography or liquid adsorption	Medium	Medium	Low	Depends	Low

Comparison of different membrane filtration means

Class	Pore size, Å	Applications
Reverse osmosis (RO)	1-10	Desalination, sugar concentration
Nanofiltration (NF)	8-50	Sugar concentration, water softening, pesticide removal, sulphate removal
Ultrafiltration (UF)	10-200	Electro-coat paint concentration, food processes, cold sterilization, waste recovery: latex &milk production
Microfiltration (MF)	200-1000	Sterilization, particle removal
Electro Dialysis (ED)	NA: Charge-oriented	Salt recovery/desalination, cheese whey purification, acid recovery

Comparison of different membrane modules

Module type	Channel spacing(cm)	Packing density (m^2/m^3)	Energy costs	Particulate plugging	Ease of cleaning
Hollow fiber	0.02-0.25	1200	Low	High	Fair
Tubular	1.0-2.5	60	High	Low	Excellent
Flat plate	0.03-0.25	300	Moderate	Moderate	Good
Spiral wound	0.03-0.1	600	Low	Very high	Poor-fair
Rotating	0.05-0.1	10	Very high	Moderate	Fair

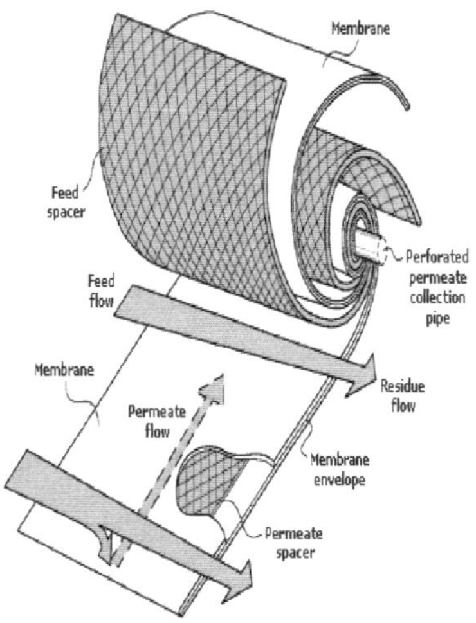

2. Equipment and procedures

Ultrafiltration: ceramic hollow fiber membrane with 200nm pore size

Process: keep 80℃ while pressure drop within 20psi. Permeate streams go for nanofiltration

Purpose: remove macromolecules (big lignocellulose fragments) and solids

Membranes and device used for nanofiltration

Parameter	DK4040f1020	AG90
manufacturer	GE-High rejection NF	GE-High rejection RO
configuration	Spiral wound	Spiral wound
Filtration area $ft^2(m^2)$	98(9.1)	90(8.4)
Average permeate flow $gpd(m^3/day)$	2,000(7.6)	2,200(8.3)
minimum salt rejection	98%	99.3%
Working pH range	3-9	4-11

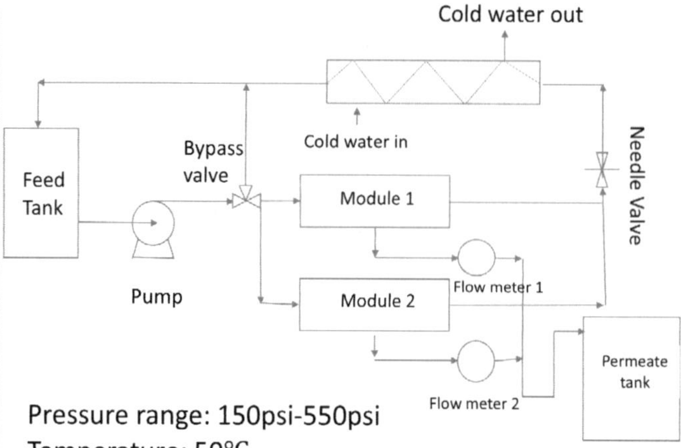

Pressure range: 150psi-550psi
Temperature: 50°C

Composition analysis

1. For feed samples, use both ^1H and ^1H-^{13}C HSQC NMR images
2. For permeate samples, use ^1H NMR image to measure inhibiters first. Then use vacuum evaporator to concentrate so that we can use ^1H-^{13}C HSQC NMR imagesto test sugar concentrations

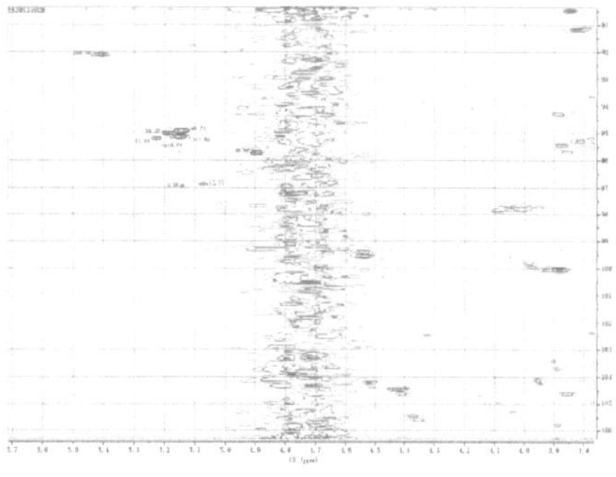

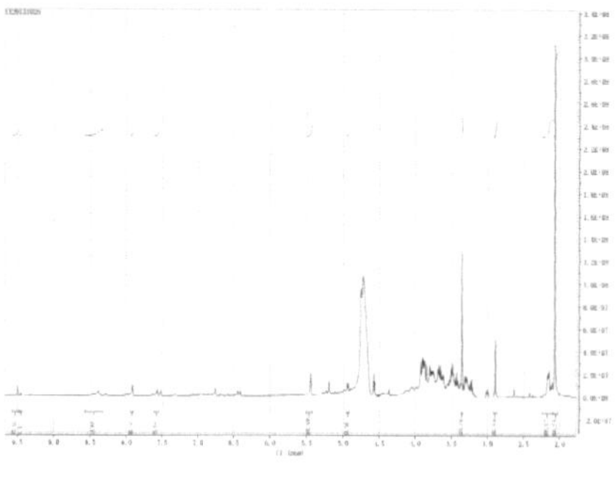

3. Mathematical model

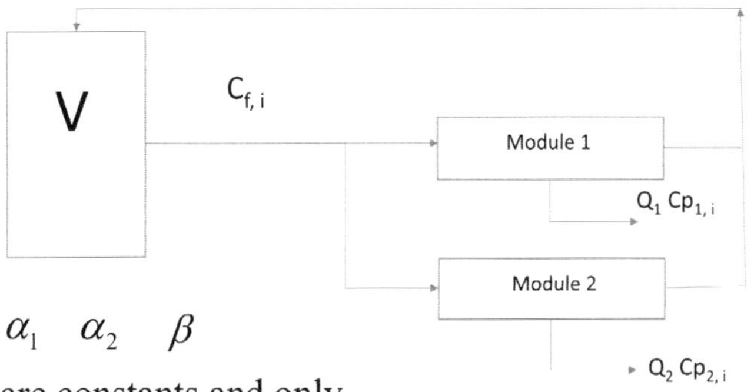

$$\frac{dV}{dt} = -(Q_1 + Q_2) \quad (1)$$

$$C_{P1,i} = \alpha_{1,i} C_F \quad (2)$$

$$C_{P2,i} = \alpha_{2,i} C_F \quad (3)$$

$$\frac{d(VC_{F,i})}{dt} = -(\alpha_{1,i}Q_1 + \alpha_{2,i}Q_2)C_{F,i} \quad (4)$$

α_1 α_2 β

are constants and only determined by membrane and solutes

$$Q_1 = \beta Q_2 \quad (5)$$

$$\frac{dV}{dt} = -(\beta + 1)Q_2 \quad (6)$$

$$\frac{d(VC_{F,i})}{dt} = -(\alpha_{1,i}\beta + \alpha_{2,i})Q_2 C_{F,i} \quad (7)$$

$$Q_2 = \frac{1}{-(\beta+1)}\frac{dV}{dt} \quad (8)$$

$$V\frac{dC_{F,i}}{dt} = -\Theta_i C_{F,i}\frac{dV}{dt} \quad (9)$$

Where $\Theta_i = 1 - \frac{\alpha_{1,i}\beta + \alpha_{2,i}}{\beta + 1}$ (10)

Solution: $C_{F,i} = C_{F0,i}(\frac{V_0}{V})^{\Theta_i}$ (11)

Retention: $R_i = 1 - \frac{C_{p,i}}{C_{f,i}}$ (12)

4. Results analysis and discussion

	Methanol	HMF	Formic acid	furfural	free acetic acid, mol/L	bound acetyl, mol/L
Final concentration g/L	0.76	1.685	1.168	0.84	0.0729	0.425
concentrated ratio	0.912	3.652	3.778	1.62	2.287	103.235
α1 (dk)	0.999	0.821	0.561	0.92	0.79	0.0166
α2 (ag)	0.971	0.163	0.397	0.586	0.419	0.0539
	Galactose	**Glucose**	**Xylose**	**Arabinose**	**Rhamnose**	**Mannose**
Final concentraion g/L	22.7	27.52	26	9.43	10.38	-
concentrated ratio	14.1	19.93	17.2	11.84	9.25	-
α1 (dk)	0.0146	0.0204	0.0276	0.0127	0.015	0.015
α2 (ag)	0.0137	0.0191	0.0165	0.0125	0.0144	0.01

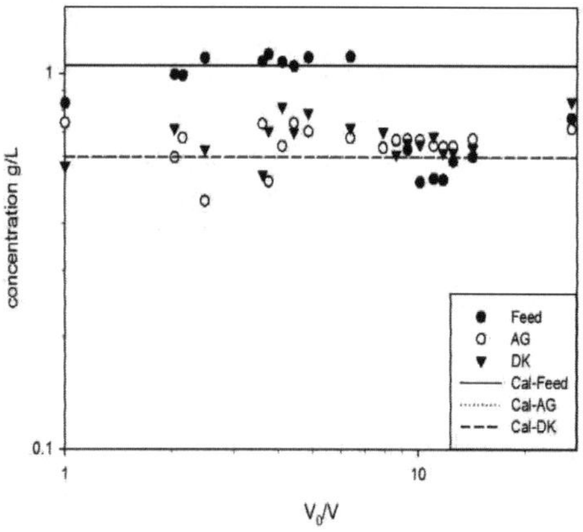

Variation of methanol concentration during NF process

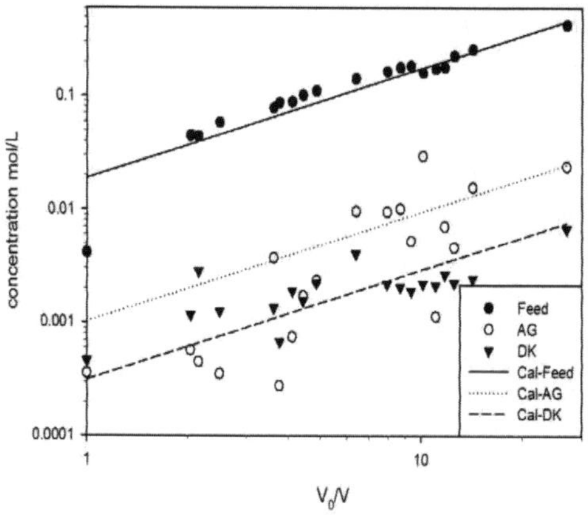

Variation of bound acetyl concentration during NF process

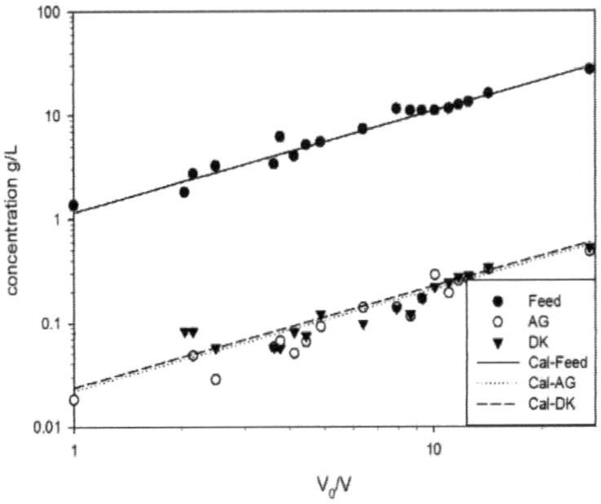

Variation of glucose concentration during NF process

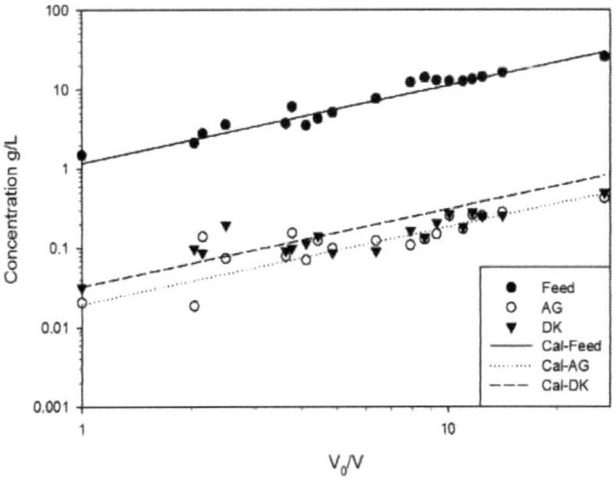

Variation of xylose concentration during NF process

5. Conclusion

1. Nanofiltration is an effective mean to remove fermentation inhibitors while preserving sugars.
2. Retentions of each species are subject to the difference of molecular weight, hydrophobicity and membrane salt rejection.
3. Nanofiltration is not efficient in separating hexoses and pentoses.

6. References

1. Shijie Liu, Thomas E. Amidon, and Christopher David Wood. Membrane filtration: concentration and purification of hydrolyzates from biomass. Journal of biobased materials and bioenergy. Vol. 2, 121-134, 2008.
2. Huajiang Huang, Shri Ramaswamy, U.W. Tschirner, B.V. Ramarao. A review of separation technologies in current and future bio-refineries. Separation and purification technology 62(2008), 1-21.
3. J. Yan, N. Joshee and S. Liu 2013 "Kinetics of the Hot-Water Extraction of Paulownia Elongata Woodchips" J. Bioproc. Eng. Biorefinery. 2(1):1-10.
4. R.D. Noble and S.A. Stern. Membrane separations technology, principles and applications. Elsevier Science B.V., New York (1995).
5. Marcel Mulder (1996). Basic Principles of Membrane technology. Kluwer Academic Publisher, Norwell, MA.
6. Fanglei Zhou, Cunwen Wang, Jiang Wei. Separation of acetic acid from monosaccharides by NF and RO membranes: Performance comparison. Journal of Membrane Science, 429(2013): 243-251.

Thank you for your attention!
Question?

373981 Design of an Organic-Inorganic Amine Monomer to Prepare Thin Film Composite Membranes for Bioethanol Purification

Abstract:

A novel hybrid organic-inorganic amine monomer has been designed and synthesized to prepare thin film composite (TFC) membranes for bioethanol dehydration via pervaporation. Different from the conventional amine monomer, m-phenylenediamine (MPD), used for interfacial polymerization, the new monomer was molecularly designed to sandwich an inorganic component nonafluorohexylmethyldichloro silane (ClSi) between two MPD molecules. Nuclear magnetic resonance (NMR) and Fourier transform infrared (FTIR) spectrometry were used to characterize and verify the chemical structures of the new monomer and the polyamide film. As a result, the advantages of polymeric and inorganic materials are effectively integrated in the newly developed TFC membranes, and these membranes exhibit a superior pervaporation separation performance as compared to most reported membranes in the literature for ethanol dehydration. Their best pervaporation performance shows a flux of 1.9 kg/m^2-h and a separation factor of 108 for the dehydration of the 85/15 wt% ethanol/water mixture at 50 °C. These novel TFC membranes show a great potential to compete with commercially available membranes.

379632 Distillation and Isolation of Commodity Chemicals from Bio-Oil Made By Tail-Gas Reactive Pyrolysis

Abstract:

Owing to instabilities, very little has been accomplished with regards to simple cost-effective separations of fast-pyrolysis bio-oil. When heated, the resulting depletion of valuable compounds (already at low concentrations) renders bio-oil distillation as a futile effort. A recently developed pyrolysis process based upon tail-gas reactive pyrolysis (TGRP) significantly improves bio-oil quality through reduction of oxygen to < 10 wt%, reduction of viscosity, increased hydrocarbon concentrations, and increased thermal stability.

We used fractional distillation to isolate compounds from bio-oil produced by TGRP. All bio-oils produced from TGRP contained significantly less acid (0.2 – 4% total), trace amounts of aldehydes, and significantly higher concentrations of hydrocarbons and phenolics (5 – 20% each). Organic yields from distillation of TGRP bio-oil ranged from 55 – 65%. On a dry basis, organic yields were more than 3 times that from traditional bio-oil; discounting acetic acid and furfural, TGRP distillation yields jumped to 10-fold. We categorized distillates according to boiling temperature and chemical components: benzene-toluene-xylenes (BTX), styrene, indene, pyridine, phenols, naphthalenes, acetic acid, and fluorene/anthracene. The relatively narrow product distributions allowed for the isolation of pure naphthalene by recrystallization from a naphthalene-rich fraction. Using GC-MS compositions, a thermodynamic computation of the TGRP distillation profiles correctly predicted the experimental results when acid content was < 1%.

Effect of operating conditions on the pervaporation of butanol from aqueous binary solutions for bio-butanol production

Hoda Azimi, Peter Gaudet, Jules Thibault, F.Handan Tezel

Handan.Tezel@uottawa.ca

Bio-butanol can be used as a renewable fuel, replacing gasoline to be used in existing car engines without modification, since it can be produced from agricultural waste. In this study, recovery of bio-butanol from acetone- butanol-ethanol (ABE) fermentation broth has been considered by using pervaporation.

A commercial PDMS (poly dimethyl siloxane) membrane was used to investigate the effect of operating conditions on performance of pervaporation for the separation of butanol from aqueous binary solutions. The effect of the feed composition and temperature on the permeation flux, separation factor and pervaporation separation index (PSI) was studied using binary solutions of butanol and water. Over the butanol concentration range of 0.5-5 wt % in the feed, the permeation flux increased with feed butanol concentration, whereas the separation factor and pervaporation separation index decreased. An increase in the operating temperature increased the permeation flux, and the increase in butanol flux was more than the associated increase in water flux, resulting in an increase in the separation factor and pervaporation separation index. Future work will include the pervaporation of ABE fermentation broths for more realistic separation applications for the production of bio-butanol.

387654 Strategies for Efficient and Cost-Effective Ionic Liquid Recycle and Product Recovery

Abstract:

Pretreatment with certain ionic liquids (IL) is highly effective at reducing the recalcitrance of a wide variety of lignocellulosic biomass to subsequent hydrolysis to release fermentable sugars. The key to an economically viable and scalable IL pretreatment technology is to develop efficient and cost competitive processes to recovery sugars and lignin and recycle IL for reuse. We have developed and tested several IL recycle and recovery strategies at the Joint BioEnergy Institute (JBEI). We have developed an improved liquid-liquid membrane extraction method that recovers >90% glucose and >95% xylose at ambient temperature and the recycled IL retains >95% of its original pretreatment efficiency. Additionally, we evaluated membrane (micro, ultra, and nano-) filtration and electro-dialysis separation technologies in conjunction with our newly developed one-pot IL pretreatment and saccharification process to enable IL recycle and recovery of lignin and fermentable sugars. Finally, we demonstrated a new regeneration protocol that use solvents precipitating cellulose but are immiscible with IL, a process that eliminates the use of water and facilitates lignin and IL recovery. Comparing with the costly distillation/chromatography, the membrane or phase separation technologies allow easier operation and less energy consumption and show great potentials for developing cost effective IL-based biomass pretreatment technologies.

386330 Separation and Purification of Biobutanol during Bioconversion of Biomass

Abstract:

Biofuels from biomass are becoming increasingly more important, due to the need for reduction in GHG emissions, energy independence, and limited global availability and increasing demand and costs of fossil fuels. Butanol has several advantages over ethanol as a drop-in bio-fuel such as higher energy content, potential for higher blending percentage with gasoline, lower vapor pressure, and lower hygroscopy. It can be used in existing transportation fuel distribution infrastructure. Butanol can be produced by fermentation of carbohydrates derived from biomass using *Clostridium acetobutylicum* or *C. beijerinckii* under anaerobic conditions. There are still many unsolved challenges for making biobutanol technically, and economically viable. The unsolved challenges lie in severe product (especially butanol) inhibition during bioprocessing, which leads to low butanol yield and productivity, and very low final product concentration (<3 wt%), causing expensive downstream processing (product separation) costs. There are two ways for solving these problems. One is the modification of microorganisms for ABE (Acetone, Butanol and Ethanol) fermentation by genetic engineering, which could keep the microorganisms alive and active under higher concentration of products in the broth. This could significantly increase the product yield, productivity, and concentration and hence reduce the production costs. However, this is still an unrealized long term goal. Another approach is the development of efficient separation and purification processes for product recovery. And, even if the modification of microorganisms becomes a reality, product separation and purification will still remain a major critical challenge. In this study, an extensive review of separation and purification of butanol from fermentation broth is provided, including gas stripping, vacuum flash, liquid-liquid extraction, membrane solvent extraction or perstraction, membrane pervaporation, membrane distillation, thermopervaporation, reverse osmosis, adsorption, and integrated bioprocessing with various separation methods. It is concluded that membrane pervaporation, solvent extraction, and adsorption are the most energy-efficient approaches for removal of butanol from the ABE fermentation broths. Integration of bioreactors with these separation methods could significantly increase the product yield, productivity and concentration and hence lower the production cost. Butanol dehydration is also discussed. This review could be helpful in the research and development and commercialization of renewable drop-in butanol biofuels and biochemicals.

362808 Revealing a Stimulatory Effect of Lignin in Enzymatic Hydrolysis of Lignocellulose

Abstract:

The electrostatic and hydrophobic interactions between cellulases and lignin play essential roles in enzymatic hydrolysis of lignocellulsoic substrates for biochemical conversion of biomass to fuels and chemicals. The objective of this study was to identify an appropriate parameter to elucidate both inhibitory and stimulatory effects of organosolv lignins on enzymatic hydrolysis. Remarkably, we observed contrasting effects of hardwood organosolv lignin (EOL-SG) and softwood organosolv lignin (EOL-LP) on enzymatic hydrolysis of Avicel and pretreated substrates. The addition of EOL-SG significantly improved the 72 h hydrolysis yields of organosolv pretreated sweetgum (OPSG) and loblolly pine (OPLP) from 49.3% to 68.6% and from 41.2% to 60.8%, respectively. On the contrary, the addition of EOL-LP (8 g/L) decreased the 72 h hydrolysis yields of OPSG and OPLP to 42.0% and 38.1%, respectively. A strong correlation between the distribution coefficients of cellulase enzymes on lignins and the changes of hydrolysis yields indicated that the inhibitory or stimulatory effects of organosolv lignins on enzymatic hydrolysis were governed by the distribution coefficients (*R*). The different *R* values of EOL-SG and EOL-LP probably were related to the electrostatic interactions, hydrophobic interactions and hydrogen bondings between enzymes and lignin, which in turn were controlled by the physical and chemical natures of lignin. Our study suggested that organosolv lignins with low affinity to enzyme could be used as additives to enhance enzymatic hydrolysis.

388143 Modeling of Swelling and Dissolution of Cellulosic Fibers for Efficient Woody Biomass Utilization

Abstract:

Dissolution of cellulose, a nanostructured polymer abundant in nature, is a critical step for the efficient utilization of this renewable resource as a starting material for the synthesis of high value-added functional polymers and chemicals and also for biofuel production. The recalcitrance of cellulose microfibrils, which is a function of complex network of cellulose, partial crystalline structure, and the extended noncovalent interactions among molecules, provides the major barrier to solubility of cellulose. There are a few solvent systems effective for direct dissolution of cellulose, and they operate under rather strict conditions of composition and temperature. Moreover, the mechanism of cellulose dissolution from solid state to homogenous solution is not well-understood. We present here a phenomenological model for describing the swelling and dissolution of solid cellulosic fibers. This model is based on the transport phenomena governing the dissolution of solid cellulose, e.g., solvent penetration, transformation from crystalline to amorphous domains, specimen swelling, and polymer chain untangling, as well as the thermodynamics and kinetics of dissolution. The model considers cellulosic fibers as non-uniform semicrystalline polymers with different properties among their cross sections due to the different supramolecular and morphological structures of cell-walls. The model predicts the: (i) crystalline, amorphous and solvent concentrations as functions of time and position within the fiber; (ii) fraction of fiber dissolved and the degree of crystallinity of the fiber with time; (iii) diameter of fiber as well as its swelling rate; and (iv) overall dissolution time. The effects of various model parameters on the kinetics of dissolution are analyzed, and the results are compared with experimental studies. The insights obtained from this model should facilitate the design of efficient solvent systems and processing conditions for woody biomass dissolution.

Abstract Submission to Be Considered As An Oral Presentation at AICHE 2014 Annual Meeting in Recalcitrance of Woody Biomass Session

Investigations into the Chemical and Structural Factors Contributing to Large Differences in the Biomass Recalcitrance of Douglas Fir Trees.

Scott Geleynse, Xiao Zhang
Washington State University

A method for screening the recalcitrance of softwood biomass toward deconstruction of cellulose was devised and applied to a group of over five hundred plantation-grown Doulgas fir trees. The trees selected exhibited considerable differences in biomass recalcitrance despite their similar origin; some samples tested exhibited fermentable carbohydrate yields of over three times difference. In this presentation, we will discuss our investigations into the chemical and structural variations in several of these trees selected to represent a cross-section of the variation in recalcitrance of Douglas fir. By examining these trees showing widely differing levels of biomass recalcitrance but from similar origin, this study aims to determine the subtle differences that contribute to the variations in biomass recalcitrance of softwood biomass.

Presenting author:
Scott Geleynse
Bioproducts Sciences and Engineering Laboratory
School of Chemical Engineering and Bioengineering, Washington State University
2710 Crimson Way, Richland, WA 99354

Contact: scott.geleynse@email.wsu.edu, x.zhang@wsu.edu

379642 Two Stage Pretreatment for Fractionation and Enhanced Enzymatic Hydrolysis of Lignocellulosic Biomass

Abstract:

Cellulosic biomass is a renewable, carbon neutral resource that can be used to generate chemicals that replace petroleum products. Optimal processes for biomass utilization separate the raw material into its constituent parts: hemicellulose, cellulose and lignin, and then offer effective strategies to utilize the chemical components of each part. Dilute acid pretreatment is an effective pretreatment for removal of the hemicellulose component but the remaining cellulose - lignin mixture is still difficult to utilize. In this work we show that a second stage of pretreatment with N-Methyl Morpholine N-Oxide (NMMO) both helps to separate the cellulose and lignin and makes the cellulose more readily digestible by cellulase enzymes. This allows for a near complete fractionation of the biomass into streams that can be easily converted into sugars and other high value chemicals.

Specific work accomplished so far has examined the treatment of corn stover and sugar cane bagasse. The corn stover was initially subjected to dilute acid treatment with a sulfuric acid solution and the bagasse was treated with phosphoric acid. The resultant material was then dissolved in NMMO solution and then reprecipitated as an amorphous material. This mass was then reacted with a commercial cellulase enzyme mixture to generate glucose. In both cases, the materials treated with both acid and NMMO, showed the same final conversion to sugars in 24 hours that took 72 hours with acid treatment alone.

In this presentation we will show the details of the treatment strategy, compare the reactivity of materials with different treatments and show the overall mass balance for biomass fractionation.

371068 Disrupting Cellulose for Conversion to Glucose in Loblolly Pine Using Ionic Liquid-Glycerol Mixtures

Abstract:

Forestry products such as Loblolly pine, containing highly crystalline cellulose and high lignin content, can be difficult to convert into the liquid biofuels our world needs to replace fossil fuels. One way to remove lignin and disrupt cellulose in lignocellulosic biomass is to use ionic liquids such as 1-ethyl-3-methylimidazolium acetate or 1-ethyl-3-methylimidazolium formate. These ionic liquids are effective in producing high cellulose residues of a less crystalline nature, which can be hydrolyzed to produce glucose that can be fermented to ethanol or other biofuels. However, ionic liquids are quite expensive. Since they are salts molten at room temperature, ionic liquids can be recycled in pine-to-glucose processing, but there are inevitable losses. Diluting ionic liquids with other "green" solvents while maintaining their efficacy is necessary for this process to be economically viable. One such environmentally-friendly solvent is glycerol, also known as glycerine. It is a low-cost (~20 cents per kg) by-product of biodiesel production, which is a growing industry. The ionic liquid 1-ethyl-3-methylimidazolium formate can be diluted half and half with glycerol to pretreat Loblolly pine for 3 hours at 140 °C and atmospheric pressure. The residue obtained gives better glucose yield from Loblolly pine after enzymatic hydrolysis than if pure ionic liquid or pure glycerol were used. More than 40% of the cellulose in the original pine was able to be converted to glucose after the 50:50 ionic liquid: glycerol pretreatment.

392329 Extracellular PHA Production from Woody Biomass

Abstract:

PHAs are commonly produced as bacterial carbon and energy storage intracellular. Recovery and purification of PHA are thus costly besides the production of cell biomass. Extracellular production of the hydroxyl acids can revolutionize the PHA production by increasing efficiency and enhancing product utility. Woody biomass can be converted to R-HAs via sugar-based platform. Hemicellulosic hydrolysates as well as cellulosic hydrolysates are excellent substrates for R-3-hydroxybutyric acid (R-3-HB) production by *Burkholderia cepacia*. Process optimization can be applied to maximize the yield and R-HA titer in the final fermentation broth.

371108 Strategies to Achieve High-Solid Enzymatic Hydrolysis of Pretreated Corn Stover

Abstract:

Enzymatic hydrolysis of pretreated biomass is one of the key steps for biofuel and biochemical production. To make the overall process economically feasible, enzymatic hydrolysis at high-solid loading is necessary since it offers many advantages over at low-solid loading, including higher sugar concentration, lower capital cost in hydrolysis, reduced a burden in separation and purification process. However, the final sugar yield decreases with the increased solid loading. Based on the preliminary result with washed diluted acid pretreated corn stover, hydrolysis at 15% WIS (water insoluble solid) with an enzyme dosage of 5 FPU/g-substrate resulted in a final sugar concentration of 94.2 g/L, which was much higher than that of 31.4 g/L obtained at 5% WIS, while the final sugar yield was decreased from 77.3% to 69.3%. To achieve both a high sugar concentration and a high final sugar yield, three different multi-stage process configurations have been proposed and tested, including (1) fed-batch, (2) splitting/thickening, and (3) clarifier processes. At 15% WIS, the final sugar yield of 76.6% was obtained from the optimized fed-batch process after 96 hr of hydrolysis. For the splitting/thickening process, the total sugar yield was higher than the one-stage process from the second cycle. We will provide detailed description on the process concepts and comparison among the three strategies.

375963 Enzymatic Hydrolysis of Waste Rejects from OCC Mills – Effect of Minerals and Optimal Methods to Reduce Inhibition

Abstract:

Fines are produced while recycling OCC (Old Corrugated Containerboards) and are usually rejected into the solid waste stream that further requires landfilling. The major component of fines rejects is primarily cellulose that can be hydrolyzed into sugars and thus present a ready cellulosic source which is a waste stream and does not require any complex pretreatment processes typical of lignocellulosic substrates such as wood chips and cereal straws or stover. However, the high content (over 30%) of ash in fines has inhibitory potential to enzymatic hydrolysis. Especially, PCC among fillers was the most powerful inhibitor reducing hydrolysis yield by adsorbing cellulase and changing pH of buffer solution. In this study, addition of anionic surfactants (SDS), cationic surfactant (CTAB) and nonionic surfactant (Tween-80) were investigated to improve enzymatic hydrolysis yield of reject fines from a recycled OCC mill. Nonionic surfactant reduced the requirement of enzyme (*Trichoderma reesei* ATCC 26921) from 50 FPU to 30 FPU for the maximum hydrolysis yield of 43% sugar (g/g-OD fines). The activity of Tween-80 reducing inhibitory potential of fillers was revealed by enzyme adsorption test and the Langmuir isotherm resulted in reduction of the affinity of fillers to enzyme. pH adjustment of buffer with acid was also studied to make a optimal hydrolysis condition for cellulase. Presence of calcium carbonate increased the pH of buffer solution with fines over 6. Addition of acid to make the buffer to pH 5 increased hydrolysis yield and combination method with nonionic surfactant reduced enzyme dosage to 20 FPU for the maximum yield. A model for the enzymatic hydrolysis of such complex mixtures was developed and applied.

Advances in Enzymes for Commercial Cellulosic Biomass Processing

Vinod Srinivasan
DuPont Industrial Biosciences
925 Page Mill Rd
Palo Alto, CA 94304
vinnie.srinivasan@dupont.com

DuPont™ Industrial Biosciences has become a world leader in industrial biotechnology. With nearly 400 enzyme products for use in industrial applications from food and beverages to household cleaning and biofuels, DuPont™ Industrial Biosciences is a pioneer in industrial enzyme innovation. Leveraging DuPont™ Genencor® Science, DuPont™ Industrial Biosciences is able to provide enzyme solutions for biomass hydrolysis and offers complete, integrated solutions along the entire biorefinery value chain. Recent advances in Accellerase® cellulosic enzyme technology, combined with DuPont™ Cellulosic Ethanol technology, has allowed DuPont™ to build its first commercial cellulosic ethanol plant in Nevada, Iowa, expected to produce 30 million gallons of cellulosic ethanol per year. The enzymatic saccharification performance results obtained using Accellerase® enzymes on various pretreated biomass feedstocks will be presented. Additionally, key learnings and feedback from customers in the emerging biomass to 2nd generation ethanol industrial sector will be discussed.

372678 The Effect of Types of Lignocellulosic Biomass on Autohydrolysis Liquor Inhibitor Composition

Abstract:

A liquor stream from autohydrolysis pretreatment of lignocellulosic biomass is enriched in dissolved hemicellulose and lignin-derived compounds, which have been demonstrated as inhibitory towards downstream enzymatic hydrolysis and fermentation. A better understanding of its inhibitory potential is sought through compositional analysis of different liquors to identify water-soluble inhibitors and their respective concentrations. Six different woody and non-woody feedstocks were hydrothermally treated at 180° C for 40 min to obtain liquors and the chemical composition was analyzed for carbohydrate content in the forms of oligomer and monomer, sugar degradation compounds (acetic acid, formic acid, furfural, 5-hydroxymethylfurfural), acid soluble and insoluble lignin, phenolic monomers, and ash. The inhibitory capability of lignin-derived phenolic monomers with respect to enzymatic hydrolysis was tested by comparison of total monomeric saccharide yields from enzymatic hydrolysis of bleached pulps enriched with pure phenolics and an inhibitor-free reference. Per 100 g of wheat straw, autohydrolysis liquors were found to contain 15.93 g of carbohydrate monomers and oligomers, 2.46 g of sugar degradation compounds, 5.47 g of lignin, 2.33 g of ash, and a low quantity of a variety of phenolic monomers. Enzymatic saccharification yields from bleached hardwood pulp decreased by 5.3% in the presence of high concentrations of the phenolic monomer vanillic acid. Liquors from woody biomass contained a lower amount of monomeric and oligomeric carbohydrates, acid soluble and insoluble lignin, and ash. Experimental work to identify and quantify phenolic monomers present in autohydrolysis liquor from different biomasses continues in effort to identify specific sources of inhibition to downstream saccharification and fermentation.

Abstract Submission to Be Considered as an Oral Presentation at AICHE 2014 Annual Meeting in T4A08 Thermochemical Conversion of Biomass I session

Potential of Alkaline Hydrogen Peroxide (AHP) as Softwood Forest Residue Pretreatment Method: Hemicellulose and Cellulose Conversion to Sugars and Chemicals

Carlos Alvarez-Vasco, Xiao Zhang

Pretreatment of softwood forest residues for sugars and biofuels production is most challenging largely due to its high lignin content and the difficulty in breaking down lignin compared to hardwood and grass. Acid or SO_2 pre-impregnated steam explosion and organosolv pretreatments have been employed. Both are facing high capital cost and low recovery of hemicellulose. Alkaline hydrogen peroxide (AHP) is an effective delignification agent and has been widely employed in lignin removal/degradation in paper industry. Recently attention has been put in investigating the use of AHP for biomass pretreatment with satisfactory results for hardwood and grasses.

This study investigated the potential of AHP for softwood forest residues pretreatment in order to maximize hemicellulose conversion to valuable chemicals and cellulose recovery to fermentable sugars. Considerable amounts of carboxylic acids including lactic acid, glycolic acid, succinic acid and formic acid were produced from hemicellulose along with pretreated biomass. The main degradation pathways involved in glucomannan conversion to organics acids were elucidated.

In addition, the role of AHP pretreatment conditions (i.e. temperature, reaction time, H_2O_2 and alkaline loadings) on resulting enzymatic cellulose-to-glucose yields of pretreated softwood were study. It was found that both temperature and alkaline dosage play an important role in peroxide decomposition-to-reactive-species which define delignification and hemicellulose depolymerization degree. The hydrophobicity of alkaline treated glucomannan and its Langmuir adsorption isotherm explain the cellulose-to-glucose yields on AHP pretreated samples as a consequence of its binding to cellulases.

This study demonstrated an integrated AHP treatment of softwood forest residues with the opportunity to produce high enzymatic cellulose-to-glucose yields from pretreated biomass and valuable chemicals from hemicellulose in order to maximize the value of biomass carbohydrates.

Presenting author: Carlos Alvarez-Vasco

Bioproducts Science Engineering Laboratory, Washington State University, 2710 Crimson Way, Voiland School of Chemical Engineering and Bioengineering, Richland, WA 99354

Contact: c.alvarezvasco@email.wsu.edu, xiaozhang@tricity.wsu.edu

384413 Comparative Techno-Economic Analysis of a Softwood Ethanol Process Featuring Post-Hydrolysis Sugars Concentration Operations and Continuous Fermentation with Cell Recycle

Abstract:

Many softwood fermentation processes are constrained by low sugars and high inhibitor concentrations in the hydrolysate. Low sugars concentrations negatively impact economics by increasing distillation costs and requiring larger equipment throughout the process, while inhibitors reduce fermentation rates, requiring larger fermentor sizes. Though high solids pretreatment and hydrolysis can improve fermentable sugar concentrations, these processes result in reduced sugar yields and increased mixing requirements. Sugars concentration operations after low solids hydrolysis are another technique to increase fermentable sugar concentrations; unfortunately, inhibitors are often concentrated along with the sugars. This analysis compares the use of multiple effect evaporation (MEE) and nanofiltration (NF) to concentrate glucose in a ponderosa pine hydrolysate from 30 g/l to 100, 150 and 200 g/l before *S. cerevisiae* fermentation. To ferment this high gravity, inhibitor containing feed, conventional batch fermentation was compared with continuous fermentation with and without microfiltration (MF) cell recycle.

Detailed capital and operating cost analyses were used to calculate the equivalent annual operating cost (EAOC = amortized capital cost + yearly operating costs) for a 5 MGY ethanol production facility. Evaporation offers the benefit of complete furfural and partial acetic acid and HMF removal, while nanofiltration has the potential to remove 65-70% of all inhibitors. Analysis showed that concentration to 100 g/l provided the most economic benefit; at higher sugars concentrations high inhibitor levels coupled with high final ethanol titers increased fermentor sizes and negatively impacted EAOC. Evaporation is more capital and energy intensive than distillation and did not reduce EAOC for any glucose level evaluated, though batch fermentor size and distillation costs are reduced substantially. Nanofiltration is not as energy intensive, but requires replacement of expensive membranes due to fouling; however, it has the potential to lower production cost by $0.10 per gallon (24% reduction in capital and 10% increase in operating expenses) when coupled with batch fermentation. Continuous fermentation with cell recycle coupled with nanofiltration was the most economical of the systems studied, reducing production cost by $0.30 per gallon (56% reduction in capital and 16% increase in operating expenses) when glucose is concentrated to 100 g/l. Sensitivity analyses investigating solids removal assumptions and nanofiltration performance show that these assumptions could be varied considerably (up to 3 times higher solids removal cost) while still lowering overall EAOC compared to the no concentration process.

384758 Continuous Ethanol Fermentation with Microfiltration Cell Recycle Applied to a Concentrated, Solids-Free Softwood Hydrolysate

Abstract:

Continuous stirred-tank fermentation with cell recycle (CSTF-CR) offers several advantages over batch fermentation when applied to second generation ethanol production. In addition to improving volumetric productivity and reducing process downtime, high cell densities quickly metabolize inhibitors such as furfural and HMF into less toxic intermediates, significantly reducing their steady-state concentrations. This is especially important in the fermentation of high gravity hydrolysates in which inhibitors are concentrated with the sugars. In the present work, batch fermentations and design of experiments methodology were used to develop a kinetic model that accounts for the inhibitory effects of acetic acid, furfural, HMF and ethanol on *S. cerevisiae* D_5A. The kinetic model was then used to predict CSTF-CR performance, and high cell retention (85-95%) continuous fermentations were used to validate the model for feeds with and without inhibitors. CSTF-CR runs with semi-defined media (100 g/l glucose + nutrients) resulted in ethanol productivities of 30 g/l/h for an uninhibited feed and >15 g/l/h for an inhibited feed containing acetic acid (4 g/l), furfural (2 g/l) and HMF (2 g/l). A productivity of 16.8 g/l/h was obtained when operating with a solids-free pine wood hydrolysate (with glucose supplemented to 100 g/l and nutrients added) at cell densities of ~60 gDW/l. Longer runs (> 120 hours) at higher cell densities were used to improve the model and study effects specific to high cell density, continuous fermentation including reduced cell viability, reduced cell yields, and adaptation of the fermenting organism. Operating conditions such as feed rate, cell retention percentage and air sparging (to maintain microaerophilic conditions) were investigated with the overall goal of improving cell viability and CSTF-CR performance.

385269 Engineering Neurospora Crassa for Improved Cellobionate Production from Cellulose

Abstract:

The conventional gluconic acid and isobutanol production process involves five steps: pretreatment, cellulase production, enzymatic hydrolysis to make sugars, aerobic fermentation to produce gluconic acid from cellulose hydrolysate, and anaerobic fermentation to produce isobutanol from sugars. Cellulase production or purchase represents a substantial portion of processing costs. A new route for converting cellulosic biomass to isobutanol and gluconate was proposed. In this proposed route, cellulase production, enzymatic hydrolysis, and aerobic fermentation will be consolidated into a single aerobic fermentation step where cellooligosaccharide aldonates are produced from pretreated cellulose by an engineered cellulolytic fungus, *Neurospora crassa.* Isobutanol and gluconate will be produced from the cellooligosaccharide aldonates in a subsequent anaerobic fermentation step. In this study, we report engineering *N. crassa* for high yield cellobionate production from cellulose, including fermentation optimization and kinetic analysis of cellulose degradation.

362910 Quantitative Structure-Activity Relationship (QSAR) Study on Carbonyl Inhibition of Microbial Fermentation

Abstract:

Biomass pretreatment generates significant amount of fermentation inhibitors from the degradation of cellulose, hemicellulose, lignin and extractives. Significant efforts have been made to identify inhibitors in biomass hydrolysates using analytical separation and identification tools including HPLC, GC/MS, LC/MS and NMR. Understanding of the potent inhibitors that have the most pronounced effects on microbial fermentation is yet to be established. Most currently identified inhibitors all have a functional carbonyl group (C=O).

Carbonyl compounds are reactive: the positive charge of carbon can initiate a nucleophilic addition between nucleophiles and electrophiles. Consequently, the inhibitory effects of carbonyl compounds are governed by their electrophilic reactivity towards biological nucleophiles. The objective of this study was to identify the specific molecular descriptors that correlate molecular structure of carbonyls to their inhibitory activity. Quantitative structure-activity relationship (QSAR) modeling was used in assessing the reactivity of carbonyls. Molecular descriptors that were studied in this work included hydrophobicity (Log *P*), dipole moment, energy of the lowest unoccupied molecular orbital (E_{LUMO}), energy of the highest occupied molecular orbital (E_{HOMO}) and electrophilicity index (ω).

The quantitative information on carbonyl inhibition was evaluated based on the glucose initial consumption rate and the final alcohol yield. We identified that E_{LUMO} is a very good global parameter to correlate the molecular structure of carbonyl compounds to their inhibitory effects. The alcohol dehydrogenase (ADH) activity in response to vanillin inhibition was determined from the batch fermentation results. Effects of acetic acid and benzoic acid on alcoholic fermentation were also studied with and without pH control. Using those data, the correlation between fermentation inhibition and carbonyl molecular descriptors were established.

384557 Acid Extraction of the Phytate from Various Corn Ethanol Coproducts

Abstract:

The bioethanol industry is rapidly developing in the U.S, with dry milling process as the major method. This process therefore produces an increasing amount of coproducts such as distillers' dried grains with solubles (DDGS). DDGS contains large amounts of important nutrients and high levels of phytate, and is widely used in animal diets of livestock industry such as dairy, beef, swine, and poultry. However, the phytate in DDGS cannot be directly digested by nonruminant animals and large quantities of phytate phosphorus (P) are discharged into the soil with the animal wastes. This can potentially cause excessive P accumulation in soil, surface and underground water bodies. On the other hand, phytic acid (PA), the acid form of phytate, has been confirmed for its beneficial effects on human health as a nutritional compound. Effective extraction of phytic acid from corn ethanol coproducts will bring revenue to the bioethanol industry by generating the more value-added product.

This study aims at optimizing PA extraction from intermediates of the corn ethanol coproduct DDGS, including concentrated distillers stillage (CDS) and wet distillers grain (WDG). PA is extracted with various HCl level (pH between 1.0 to 4.0), at different temperatures (35 and 50 ^{o}C), extraction time (1 to 6 h), and with different ratios of liquor to solids (2:1 to 10:1). The extracted PA concentrations were determined via spectrophotometry method (ferric ion and sulfosalicylic acid) after purification by the anion exchange resin to remove heteroion such as phosphate ion. The optimization study would give the best reaction conditions for PA extraction in terms of pH, extraction time, temperature, ratio of liquor to solids.

When extraction condition optimized, a 1-L scale reactor was designed and used to recover PA from the extracted solution. The pH value of the solution was adjusted to 7.0 by adding NaOH, and phytate was then precipitated with $Ca(OH)_2$ as calcium phytate solid by sedimentation, centrifuge, and oven-drying. The whole process produced about 5 to 10 g of phytate when undiluted CDS used as feedstock.

Fragmentation of organosolv lignin by Fenton reaction facilitates production of lignin derived mucolic acid and aromatic compounds.

Jijiao Zeng[a], Changgeun Yoo, Xuejun Pan, Wilfred Vermerris

Fei Wang, Zhaohui Tong[a*]

[a] Department of Agriculture & Biological Engineering, IFAS, University of Florida, Gainesville, FL 32611, USA

* Corresponding author: Tel.:+01-3523921864; fax: +01-3523924902. E-mail address: ztong@ufl.edu (Z. Tong).

Abstract: Valorization of lignin has recently drawn a great of attentions in biorefinery process. One of alternative pathways is disruption of complicate lignin into monomolecular by either reductive or oxidative reaction. However, lignin depolymerization process is challenging because functionality and complex structure of lignin hinders external attack and usually requires harsh reaction conditions. It is know that, the transition metals (like Fe^{2+}, Mn^{2+} and Cu^{2+} ions) can reduce H_2O_2 to generate free •OH which can oxidize almost all organic chemicals in mild condition. We propose that Fenton reaction could modify and fragment of lignin which results in increasing the conversion rate of lignin in subsequent liquefaction reaction. Therefore, in this study, organosolv lignin was pretreated with Fe^{3+}/ H_2O_2 (Fenton catalyst) in different levels by adding gradient enhanced H_2O_2 concentration. The oxidation of α-OH and γ-OH to corresponding aldehydes on lignin side chain were observed by FTIR, XPS and HSQC NMR analysis. Interestingly, Fenton reaction pretreatment also caused fragmentation of lignin together with side chain oxidation. Up to 60% w/w lignin transferred into water soluble phase in highest reaction condition. The molecular weight of soluble and insoluble lignin was determined by GPC. The overall conversion rate of Fenton treated lignin was tested by subsequent supercritical liquefaction. The results showed that lignin was liquefied into high yield of mucolic acid like compounds (relatively 22% of total oil yield) and various aromatic monomers. The total liquid yield was increased from 50% to 60% w/w, while the char yield was decreased from 30% to 20% w/w. The mechanism has been ascribed to Fenton reaction hydroxylates organosolv lignin to form a Fe^{2+} or Fe^{3+}–lignin complex which greatly prohibits intramolecular interaction and reduces the chance of arrangement of $-OCH_3$.

2014 AIChE Annual Meeting
International Congress on Energy (ICE)
T4A05 Biochemical Conversion Processes in Forest/Plant Biomass Biorefineries I
November 16-21, 2014
Atlanta, GA

Pd-catalyst assisted organosolv process to isolate ethanol organosolv lignin retaining compatible characteristics for producing lignin monomeric compounds

Ho-Yong Kim[1*], Sunkyu Park[1], In-Gyu Choi[2, 3]

1 Department of Forest Biomaterials, North Carolina State University, Raleigh, NC 27695, USA
2 Department of Forest Sciences, Seoul National University, Seoul, 151-921, Korea
3 Research Institute for Agriculture and Life Sciences, Seoul National University, Seoul, 151-921, Korea

* Presenting author: hkim36@ncsu.edu

Notwithstanding many efforts to produce reliable lignocellulosic bioethanol, several critical barriers were still remained unsolved. One of the natural barriers is existence of lignin. Lignocellulosic biomass contains up to 30% of lignin, however it cannot be utilized to produce bioethanol, and it is rather inhibiting the enzymatic hydrolysis of cellulosic fibers. Thus recently, plenty of researches have been conducted to isolate lignin from the cellulosic fiber during the pretreatment. However most of isolated lignin during the pretreatment has low value due to high degree of condensation. Therefore, we applied palladium catalyst to the organosolv pretreatment process. Palladium is mainly researched in pyrolysis process to encourage hydrogenation, thus we expect to hinder the condensation of lignin fragments and produce more ethanol organosolv lignin (EOL). Sugar production was slightly declined as pretreatment temperature increased from 145°C to 160°C due to decomposition of glucan. However palladium treatment mitigated glucan decomposition, and above all EOL yield was significantly enhanced by palladium treatment, especially at 160°C (from 13.36 g to 16.30 g). Furthermore, it prevents cleavage of ether linkages and condensation reaction of lignin fragments. High yield and ether linkage rich EOL released significantly larger amount of lignin monomeric compounds. After pyrolysis-GC/MS analysis, released amount of syringol was significantly decreased while pretreatment temperature increased (from 25.8 mg/g to 6.1 mg/g), however it was remarkably recovered when palladium was introduced (from 6.1 mg/g to 29.3 mg/g).

An Efficient Approach to Solely Biomass-Derived *p*-Xylene and Terephthalic Acid

Fei Wang, Zhaohui Tong,*

Agricultural and Biological Engineering Department, University of Florida,
Rogers hall, Museum Road, PO Box 110570 Gainesville, FL. 32611

*ztong@ufl.edu

Abstract: We report a novel method to produce *p*-xylene (PX) and terephthalic acid (TPA) from biomass-derivative chemicals (isoprene and acrolein). *p*-xylene (PX) is conventionally produced from petroleum and considered as a basic chemical with high worldwide consumption. Particularly, it has been largely oxidized for the preparation of terephthalic acid (TPA), an important monomer for synthesizing polyesters including polyethylene terephthalate (PET), polypropanediol terephthalate (PPT) or polybutylene terephthalate (PBT), which have broad applications from drink bottles to polyester fibers. In this work, an important intermediate, 4-methyl-cyclohex-3-enecarbaldehyde, was prepared in a 75% yield through Diels-Alder reaction between isoprene and acrolein in presence of Lewis acid (zinc chloride) as the catalyst. Then, *p*-Xylene was obtained through Wolff—Kishner reduction followed by a dehydro-aromatization conversion from this intermediate. By sharing the same intermediate, TPA can be directly obtained in a high yield via an aromatization reaction using $CuBr_2$ as the oxidant in presence of LiBr followed by an oxidation reaction using industrial-available Cobalt (II) acetate (0.5 mol%) and Magnesium acetate (0.5 mol%) as the catalysts in the presence of N-Hydroxyphthalimide. Finally, we concluded that we have successfully prepared *p*-xylene and TPA in a high yield from bio-based isoprene and acrolein under mild conditions and using flexible routes.

386535 Lipid Production from Lignocellulosic Biomass Via a Newly Isolated Oleaginous Fungal Strain

Abstract:

Microbial lipids can be an attractive feedstock for biodiesel production to replace the current plant-oil based biodiesel. Oleaginous fungi have significant advantages, such as high growth rate and high oil production, easy harvest, good tolerance on growth conditions (e.g., inhibitors, pH), and can be easily grown and controlled in bioreactors. Several oleaginous yeast and fungal strains are capable of accumulating high amount of lipids inside cells, but rarely can directly utilize lignocellulosic biomass as substrates since the lack of cellulase producing capability. A newly isolated *Fusarium* strain from the endophytic fungi of soybean plants was discovered to generate high content of lipids (53.6-57.8% of dry biomass) and also a significant amount of cellulase enzyme compared to well-studied oleaginous strain *Mortierella isabellina*. This research focus on the growth behavior and characteristics of this new strain and the developing of different bioconversion processes utilizing lignocellulosic biomass for improved microbial lipid accumulation. The strain showed good carbon source utilization since it can produce a significant amount of lipids on most of carbon sources including glucose, xylose, mannose, galactose, glycerol, lactose, fructose, sucrose, cellobiose, starch, xylan, and cellulose. The strain cultivation (growth medium, temperature, agitation, light effect, etc.) was optimized and the extracellular cellulase production of this strain was enhanced by directed evolution. Investigation on the effect of pretreatment and cellulase enzyme were also conducted to achieve higher efficiency of converting lignocellulosic biomass to lipids.

381194 Study of the Effect of Alkali Distillation Liquid on Simultaneous Saccharification and Co Fermentation (SSCF) of Alkaline Pretreated Solid State Fermented Sweet Sorghum Bagasse with Zymomonas Mobilis

Abstract:

Sweet sorghum is a promising energy crop to produce bioethanol. An integrated process is necessary to reduce the product cost and increase the utility efficient of feedstock. An alkaline pretratment combined distillation based on the advanced solid-state fermentation technology (ASSF) has been designed to improve the bioconversion of lignocellulose into fermentable sugar. After alkaline distillation step, most of lignin moved into aqueous state from lignocellulosic substrates. Effects of these alkali liquid on the performance of enzymatic saccharification and sugars fermentation of SS with *Zymomonasmobilis*is required to be investigated because it is important to decide the following process.

In this study, effects of lignin content, ionic strength, pH and washing operation on enzymatic saccharification and fermentation were studied. The results showed that: 1) The free lignin content has no obvious negative effect on enzymatic hydrolysis. 2) Ionic strength has no effects on enzymatic hydrolysis, but high ionic strength inhibits the fermentation of *Zymomonasmobilis.* 3) Final ethanol yield was highly dependent on the monomer sugar yields during prehydrolysis step in the present of various alkali distillation liquid concentration. 4) Highest ethanol yield was achieved when the solid from solid-liquid separation was washed once.

2014 AIChE Annual Meeting
T4 International Congress on Energy (ICE)
T4A07 Biomass Conversion: Processing of Solids
November 16-21, 2014
Atlanta, GA

Multi-Scale Investigation for the Effect of Mechanical Refining for Pretreated Lignocellulosic Biomass on the Subsequent Enzymatic Hydrolysis

Junyeong Park [1,*], Wenhui Geng [1,2], Kabindra Kafle [3], Brandon Jones [1], Seong H. Kim [3], Sunkyu Park [1]

[1] Department of Forest Biomaterials, North Carolina State University, Raleigh, NC 27695
[2] Jiangsu Provincial Key Lab of Pulp and Paper Science and Technology, Nanjing Forestry University, Nanjing, China
[3] Department of Chemical Engineering, Pennsylvania State University, University Park, PA 16802
* Presenting author: jpark7@ncsu.edu

Mechanical refining, an established technique widely used in the pulp and paper industry, has recently received significant attention from industries due to its potentials as a promising process for pretreated lignocellulosic biomass. Several studies have been reported about the effects of mechanical refining on the subsequent enzymatic hydrolysis, such as increased yield of carbohydrate conversion, reduced dosage of cellulolytic enzymes to get target yields, and reduced pretreatment severity that make fermentation or catalytic upgrading more effective. In addition, it is commercially proven technology at process flows of ~1,500 dry tons/day of biomass. The details about how refining affects the enzymatic hydrolysis, however, have not been fully elucidated so far. In this study, the changes after mechanical refining of pretreated biomass were investigated by using various tools with different observation scales. Water retention value (WRV) of unrefined and refined samples, which can represent the changes in surface area and pore volume of the biomass, showed close correlations with increased yields of carbohydrate conversion. Optical microscopy and fiber quality analysis (FQA) can also support this observation. TEM provided the evidence of structural changes in nano scale, including delamination of cell wall and external fibrillation. It was also expected that the refining process will affect the molecular structure of pretreated biomass, such as crystallinity of celluloses. XRD solid-state NMR, and SFG spectroscopies were carried out to observe alterations in molecular-scale level, but no change in cellulose crystallinity was observed by those tools. Based on the results, the effect of refining on enzymatic hydrolysis seems to be mainly from the structural modification from the nano to the micron level. More details about this observation with varying factors in conversion yields, like type of pretreatment, refining severity, enzymatic hydrolysis condition, will be discussed.

376760 Effect of Temperature on Pelletizing SPORL Pretreated Forest Biomass and Subsequent Enzymatic Saccharification and Fermentation

Abstract:

Pelletizing of pretreated lignocelluloses to facilitate biomass supply chain logistics was simulated using hot-pressing. The effects of fiber hornification induced by hot-pressing on enzymatic digestibilities of lodgepople pine and poplar NE222 wood chips pretreated by Sulfite pretreatment to overcome recalcitrance of lignocelluloses (SPROL) were examined. The pretreated wood chips were pressed at 25°C, 70°C, 90°C, 110°C, and 177°C. The cellulose accessibilities of the pressed and unpressed substrates were evaluated using water retention value and direct cellulase adsorption measurements. Hot-pressing below 110°C produced a degree of hornification (DH) below 0.26 and had limited effect on cellulose accessibility and enzymatic digestibility. Hot-pressing at 177 °C produced a DH of 0.86 that significantly hornified the fibers and resulted near zero saccharification. The saccharification results were consistent with cellulose accessibility data. Ethanol fermentation studies at 18% solids suggest that a pressing below 110°C is preferred to reduce its effect on biofuel yield.

382102 Optimization of Operating Conditions of Aerobic Digester for Effective Conversion of Biomass to Bio-Gas

Abstract:

Pakistan is facing energy crises from last 20 years and share of the bio-fuel in the energy mix is also negligible. According to vision 2020 plan of Pakistan, share of alternative resources/ renewable will be 2,000MW out of 20,000 MW. Various biomasses are available but Conversion of biomass to bio-fuel at low cost is a challenge.

This paper investigates the yield of bio-gas from biomass (like molasses, fruit (banana, apple, water melon)/vegetable wastes, Fruit/vegetable peels, lignocelluloses material and dairy waste etc). Anaerobic digester (LS-26 102) is used to measure the bio-gas yield and concentration at various applied conditions like temperature, residence (time) and behavior of pH value and conductivity is recorded. Enhanced yield of bio-gas is recorded at 40 ^{0}C than 22 ^{0}C at similar other applied condition. An experimental investigation shows that conversion of all biomass at 40 ^{0}C for bio-gas formation is much higher than at 22 ^{0}C. Biogas yield for fruit and vegetable wastes is about 30-35% higher than peels of same fruits and vegetables. Increasing trend of conductivity is observed over the time which also confirms the change in Ph value (decreasing trend) and is according to the literature [1, 2]. Moreover yield of bio-gas at 40 ^{0}C, from molasses, fruit /vegetable waste, and peels of fruit / vegetable is compared. Results reveal that molasses produce higher volume of bio-gas than wastes of fruits/vegetable and peels etc under similar applied condition.

Regional potential of province of Punjab Pakistan is calculated for bio-gas from biomasses and 1500MW can be generated to achieve the vision 2020-25.

[1] Sagagi,B. S., B. Garba and N. S. Usman Bayero, Journal of Pure and Applied Sciences, 2(1): 115 – 118 Received: October, 2008 Accepted: February, 2009, STUDIES ON BIOGAS PRODUCTION FROM FRUITS AND VEGETABLE WASTE

[2]http://www.unigraphinternational.com/en/pdf_infotech/pH_and_conductivity.pd

Abstract:

Bio-based chemicals constitute a promising alternative platform for industrial chemical production, allowing for feedstock diversification, utilization of renewable resources, and potentially reduced environmental burdens relative to petrochemicals. Bio-based chemicals also present new market opportunities in a range of end-use applications, from coatings to polymers. Global production of chemicals and polymers from petroleum is about 330 million tons, and it mainly consists of basic platform chemical that are basis for production of fine chemical and polymers. Studies have shown that most of these building blocks can be produced from biomass resources.[1] Current bio-based chemical production, from both woody and herbaceous resources, is estimated to be around 50 million tons.[1] Among all available sources, lignin with annual production of 200 million metric tons/yr. is the second most abundant natural polymer on earth after cellulose.[2, 3] Lignin accounts for approximately 25–35 % of the organic matrix of wood. The monomer structure has been reported to consist of a phenyl propanoid skeleton.[4] Transformations of lignin have been developed with the goal of preserving the phenolic structure for more ready production of complex aromatic chemicals , but these routes have been difficult and relatively energy consumptive.[4, 5] There are several transformation methods such as pyrolysis and liquefaction techniques, acidic or basic hydrolysis, and more selective depolymerization through catalytic processes,[5, 6] but there is little modeling work on the economic and environmental feasibility of large-scale bio-based chemical production via these routes.

Here we consider bio-based production of catechols from catalytic depolymerization of lignin from candlenut nutshells. These nutshells have 12% (wt. %) lignin in their structure[5] and are typically burned or piled as waste. Target product, catechol, is an organic compound with molecular formula $C_6H_4(OH)_2$ consumed in the production of pesticides, the remainder being used as a precursor to fine chemicals such as perfumes and pharmaceuticals. Integrated process consists of nutshell preparation (rinsing, drying and milling) and lignin extraction with methanol followed by lignin depolymerization in presence of Cu-doped porous metal oxide (Cu-PMO) catalyst. Our study is mainly focused on catalyst synthesis and lignin depolymerization.

A life cycle assessment (LCA) was conducted in order to model predicted energy use and environmental impacts associated with this bio-based production scheme at industrial scales and to compare catechol production from both lignin and petroleum sources. All major upstream processes were considered, including candlenut cultivation, harvesting, and transport, for the bio-based route and oil&gas extraction, transport, and petrochemical refining for the petrochemical route. Lignin extraction and modification steps were modeled by scaling up bench-scale process and estimating energy consumption. Conde-Mejía, C., et al. [7] was modified based on solvent to wood ratio and used. For catalyst synthesis and lignin depolymerization, commercial-scale process simulation was performed using ASPEN plus v.8.4. Input chemicals from original research [5, 8] and data from the NREL biomass ASPEN data bank [9] were used to model the reaction components and kinetics. Simulations were based on input lignin flow rate of 75 tons/day (matching current US industrial scales for lignin processing).

Catalyst degradation (5000 hr lifetime [10]) and solvent recycling rates of 97% were assumed. Figures 1 and 2 show the ASPEN process flow sheets for these steps.

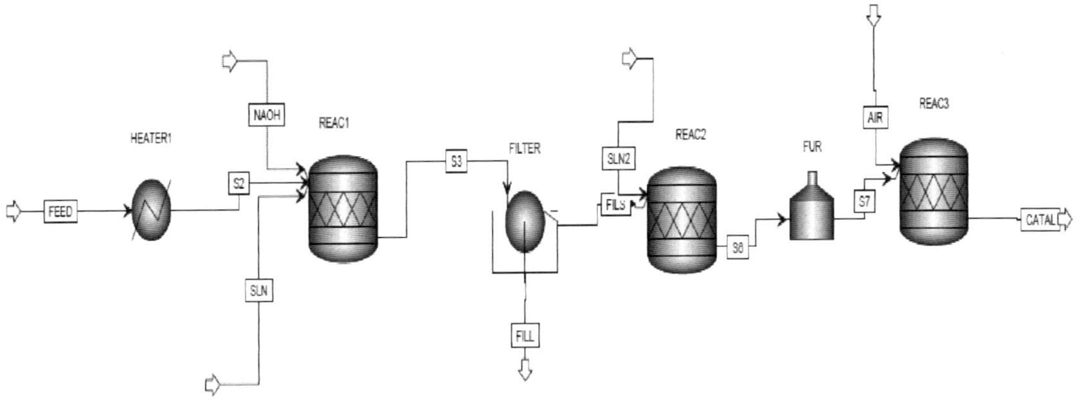

Figure 1. Flow diagram of catalyst synthesis

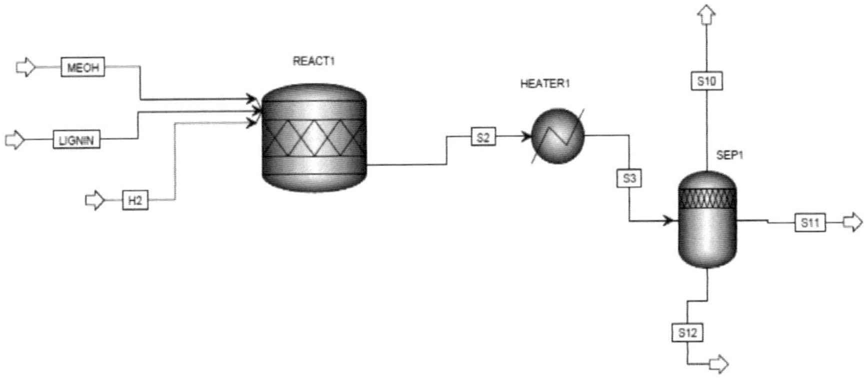

Figure 2. Flow diagram of lignin depolymerization

Energy use for catalyst synthesis is estimated as 5,920 kWh/kg of Cu-PMO catalyst, while depolymerization requires 2.46 kWh/kg of catechol. Phenol and ethanol are co-products of depolymerization process with daily production of 10 and 27 tons. Ethanol will be recycled back to extraction and used as solvent and phenol can be sold to the market.

Resource inputs (energy and material) were modeled in SimaPro 8.01 LCA software using ecoinvent 2.2 life cycle inventory unit processes adjusted for the US energy system. The U.S. EPA's Tool for the Reduction of Chemical and Other Environmental Impacts (TRACI) 2.1 fate-effect model was used for estimation of environmental impacts. Associated impacts were normalized based on 1 kg catechol in output stream, considering burden allocation for catalyst synthesis over its lifetime and two types of credits; one from carbon sequestration during candlenut tree growth prior to harvesting and the second from phenol as a co-product of depolymerization.

Figure 3 demonstrates the relative process contribution, normalized in each category, for the field-to-gate production of catechols from candlenut lignin.

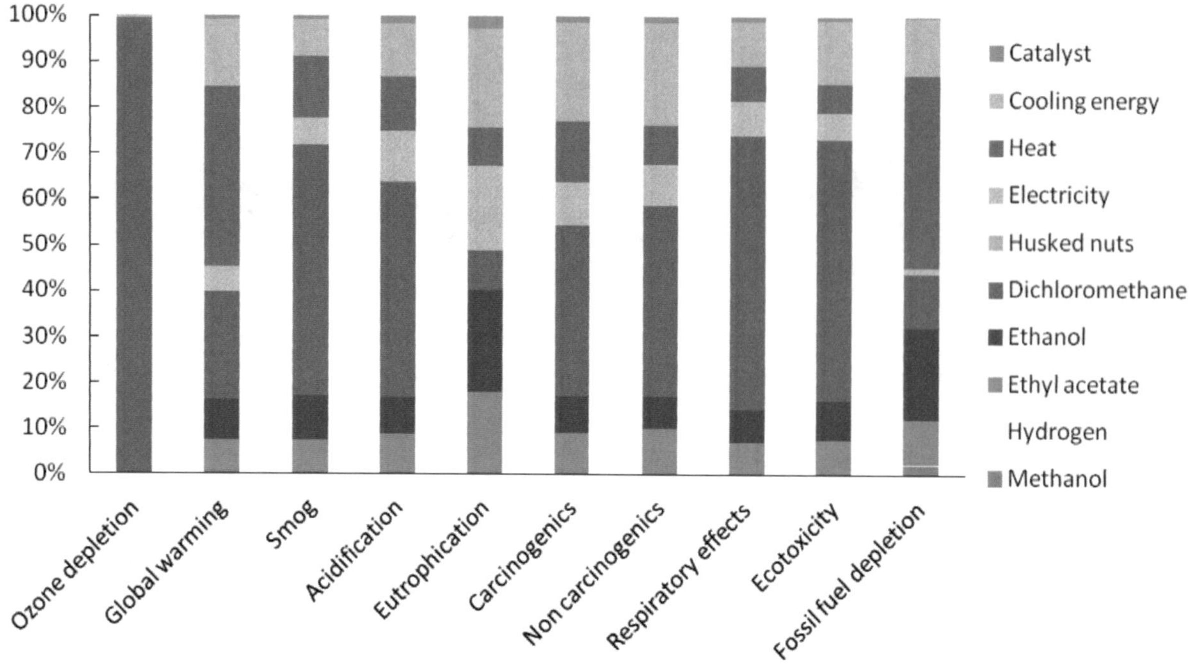

Figure 3. Process contribution of catechol production from lignin, considering extraction, depolymerization and catalyst synthesis

From the chart, it can be seen that almost in all categories, dichloromethane and heat used in extraction process drive life cycle impacts. Dichloromethane is produced from methyl chloride at high temperatures (400-500 °C) and it is highly volatile, which makes it an inhalation and atmospheric hazard. It is a potent ozone-depleting substance and its contribution is 97% in ozone depletion and 62% in human respiratory effects.

Catalyst contribution is not that significant compared to other processes. The production of 1 kg catechol requires 0.76 kg of catalyst. Results demonstrated that maintaining operational conditions for high retention time of catalyst synthesis process leads to excessive electricity consumption. Electricity production is the dominant contributor in all categories with about >99% of total burdens, while the actual catalyst materials themselves are relatively insignificant.

Final step is comparing environmental burden of proposed catechol production with convenient catechol from petrochemical phenols. Phenol reacts with hydrogen peroxide to form catechol based on equation 1.[11]

$$C_6H_5OH + H_2O_2 \longrightarrow C_6H_4(OH)_2 + H_2O\ ,\ \Delta H_{reaction} = -200.4\ (kJ/mol) \quad \textbf{eq. 1}$$

The process was simulated in SimaPro and associated burden was estimated. Its process contribution is shown in figure 4.

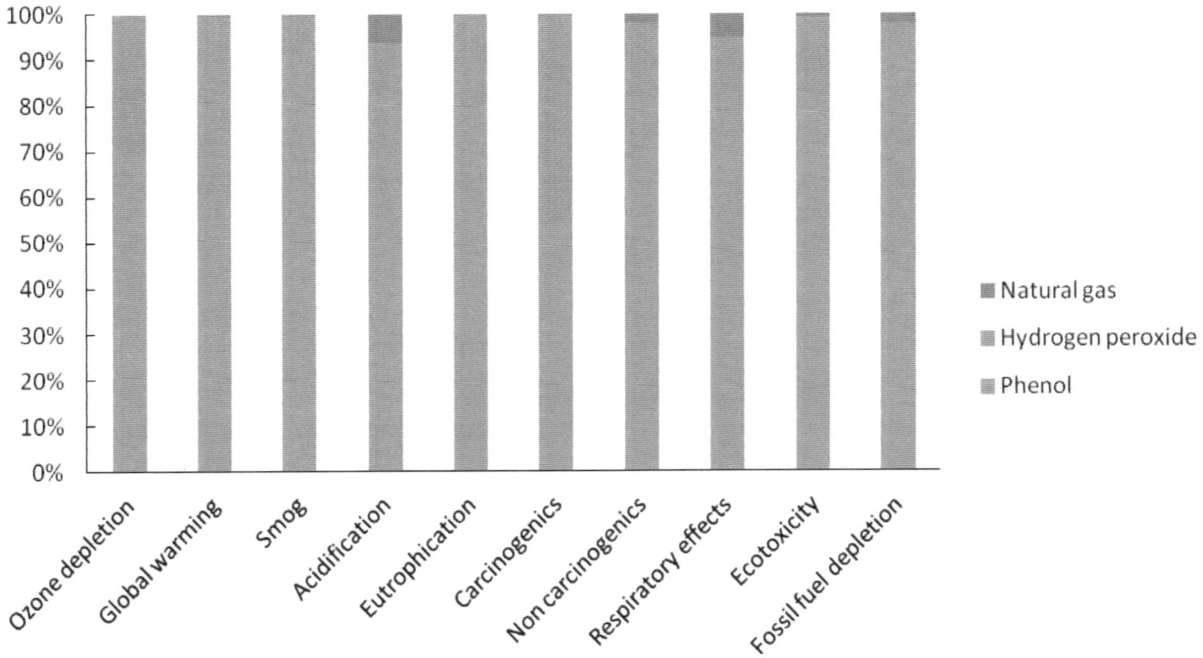

Figure 4. Process contribution in catechol production from petroleum based phenol

Based on above figure, phenol production at first place is the most dominant contributor almost in all environmental impact categories.

Comparing petroleum-based and lignin-based catechol illustrates interesting results. Table 1 shows total impacts for each of the production routs.

Impact category	Unit	Total (lignin-based catechol)	Total (petroleum-based catechol)
Ozone depletion	kg CFC-11 eq	1.6E-04	1.5E-07
Global warming	kg CO2 eq	3.0E+01	4.0E+00
Smog	kg O3 eq	1.5E+00	1.7E-01
Acidification	kg SO2 eq	1.1E-01	1.6E-02
Eutrophication	kg N eq	3.6E-02	8.8E-03
Carcinogenics	CTUh	1.0E-06	2.4E-07
Non carcinogenics	CTUh	2.5E-06	3.0E-07
Respiratory effects	kg PM2.5 eq	1.0E-02	1.2E-03
Ecotoxicity	CTUe	2.7E+01	8.5E+00
Fossil fuel depletion	MJ surplus	6.9E+01	1.3E+01

Table 1. Total environmental burden of lignin-based and petroleum-based catechol

Petroleum-based catechol seems to have lower environmental impacts compared to the bio-based catechol and that might be because of high complexity and resistance in lignin structure that needs more effort for conversion to simpler aromatic compounds. Utilization of dichloromethane as a solvent for modification of crude lignin is a driver of negative environmental impacts and its substitution should be a target of further research.

Finally, the results points that proposed pathway for bio-based catechol is not as promising as expected. More development and process optimizations are needed for industrial scale production of such chemicals. Choosing appropriate solvents and reaction modifications can be the first step while using catalyzed reactions to mitigate energy consumption of the whole process can be another consideration.

References:

1. de Jong, E., *Bio-based Chemicals Value Added Products from Biorefineries.* IEA Bioenergy, Task42 Biorefinery, 2012.
2. Ragauskas, A.J., et al., *The path forward for biofuels and biomaterials.* Science, 2006. **311**(5760): p. 484-489.
3. Percival Zhang, Y.-H., M.E. Himmel, and J.R. Mielenz, *Outlook for cellulase improvement: screening and selection strategies.* Biotechnology advances, 2006. **24**(5): p. 452-481.
4. Kleinert, M. and T. Barth, *Phenols from lignin.* Chemical engineering & technology, 2008. **31**(5): p. 736-745.
5. Barta, K., et al., *Depolymerization of organosolv lignin to aromatic compounds over Cu-doped porous metal oxides.* Green Chemistry, 2014. **16**(1): p. 191-196.
6. Huber, G.W., S. Iborra, and A. Corma, *Synthesis of transportation fuels from biomass: chemistry, catalysts, and engineering.* Chemical reviews, 2006. **106**(9): p. 4044-4098.
7. Conde-Mejía, C., A. Jiménez-Gutiérrez, and M. El-Halwagi, *A comparison of pretreatment methods for bioethanol production from lignocellulosic materials.* Process safety and environmental protection, 2012. **90**(3): p. 189-202.
8. Hansen, T.S., et al., *One-pot reduction of 5-hydroxymethylfurfural via hydrogen transfer from supercritical methanol.* Green Chemistry, 2012. **14**(9): p. 2457-2461.
9. Wooley, R.J. and V. Putsche, *Development of an ASPEN PLUS physical property database for biofuels components.* 1996: National Renewable Energy Laboratory Golden, CO.
10. Griffiths, O.G., et al., *Using life cycle assessment to measure the environmental performance of catalysts and directing research in the conversion of CO 2 into commodity chemicals: a look at the potential for fuels from 'thin-air'.* RSC Advances, 2013. **3**(30): p. 12244-12254.
11. Sienel, G., R. Rieth, and K. Rowbottom, *Ullmann's encyclopedia of industrial chemistry.* Wiley-VCH, Weinheim, 2000.

Calcium Pretreatment and the Resulting Compositional Changes in Pyrolysis Oils

Paige A. Case, M. Clayton Wheeler and William J. DeSisto

Department of Chemical and Biological Engineering, University of Maine, Orono, ME 04469

Conventional fast pyrolysis oils have many properties that make them unsuitable as direct replacements for petroleum transportation fuels, due in large part to their high oxygen content. Research efforts to overcome this issue have focused largely on catalytic fast pyrolysis with zeolites and hydrodeoxygenation (HDO) of the condensed oils. Our research has focused on applying pre-treatment with various calcium compounds to fast pyrolysis of pine. We find that some calcium compounds, including the formate salt, oxide, hydroxide and carbonate have an effect on the product yield and types of compounds produced during fast pyrolysis. We observe an increase in phenol and alkylated phenols accompanied by a decrease in catechol type compounds. There is also suppression of anhydrosugars from the hemicelluloses and celluloses. The calcium-pretreated oils are very stable during aging and can be upgraded using conventional hydrotreating catalysts under relatively mild conditions.

In this work, we probe the mechanisms that lead to the chemical changes observed with the calcium pretreatment. Specifically, we investigate the shift from catechol-type to phenol-type compounds using bench-scale pyrolysis experiments and analytical techniques such as quantitative GC-MS and carbon and hydrogen NMR. Whole biomass, as well as model compounds, is used to understand the contributions of individual components and their interactions with each other.

Woody residues represent a lower cost option for feedstocks for the production of biofuels. In this study, the pyrolysis of woody residues was investigated as part of Clean Energy Dialogue between the U.S. and Canada. Three pine-based hog fuels from saw mills and wood from pine beetle killed trees were chosen as the woody residue feedstocks and pine and birch as the reference clean feedstocks. The yields and quality of the oil were evaluated in a bubbling fluidized bed reactor and a laboratory-scale pyrolyzer connected to a molecular beam mass spectrometer.

The presence of mineral matter, in particular alkali metals, reduced the yields of bio-oil from the hog fuel feeds when compared to the reference pine. Lower yields of bio-oil from pyrolysis corresponded to higher yields of char. The presence of alkali metals also increased the ratio of CO_2 to CO in the gas and gave oils with lower oxygen contents and lower acid contents. Multivariate analysis of the laboratory MBMS data revealed further differences between the oils.

In the laboratory experiments, the total ion count for vapor species was used as a surrogate for oil yields. Good qualitative correlation between the vapor species total ion count and the oil yield in the fluidized bed experiments was obtained, which indicated that the micro-scale experiments can be used as a rapid screening tool to asses feedstocks.

We further evaluated water washing of the hog fuels as a method to reduce mineral matter and increase yields of pyrolysis oils. Water washing effectively reduced Na and Cl contents in hog fuels with high salt contents and restored oil yields to values comparable to other hog fuels.

Effects of biomass inorganic elements on catalytic fast pyrolysis over HZSM-5

Charles A. Mullen*, Akwasi A. Boateng

Eastern Regional Research Center, Agricultural Research Service, U. S. Department of Agriculture, 600 E. Mermaid Lane, Wyndmoor, PA 19038.
*charles.mullen@ars.usda.gov

Fast pyrolysis methods are effective for liquefaction of lignocellulosic biomass. Catalytic fast pyrolysis (CFP) can promote the rejection of problematic oxygen leading to the production pyrolysis oils with reduced oxygen content compared with non-catalytic methods. CFP can be performed either *in situ*, where the pyrolysis occurs over the catalyst, or *ex situ* where the pyrolysis and catalytic conversion steps are decoupled. Catalyst deactivation due to carbon formation has been well documented for CFP, but studies on the effects of inorganic components of biomass on CFP have been limited. We recently studied the accumulation of inorganic species on HZSM-5 during *in situ* CFP where pilot scale fluidized bed experiments were performed on switchgrass to track the levels of Ca, Cu, Fe, K, Mg, Na and P over HZSM-5. We found the accumulation of these elements on HZSM-5 was rapid with an observed concurrent decrease in the yield of aromatic hydrocarbons indicating deactivation of the catalyst even as carbon deposits were eliminated by intermittent combustion. It is expected that *ex situ* rather than *in-situ* processes will slow this deactivation by removing many inorganic species with bio-char prior to the catalytic step; however, the true effect of inorganic species on the incipient CFP process is still unknown. To elucidate these factors, a high throughput py-GC/MS experiments were performed on the HZSM-5 catalyzed pyrolysis of biomass samples containing different concentrations of various inorganic elements to study their effect on product yields and selectivities. A strong negative correlation was found between the yields of alkyl benzenes and biomass potassium content. The relationships between the use of *in situ* or *ex situ* catalytic pyrolysis and biomass inorganic species on catalyst deactivation, aromatic hydrocarbon yields and selectivities will be discussed in this presentation.

2014 AIChE Annual Meeting
International Congress on Energy (ICE)
T4A11 Thermochemical Conversion of Biomass II: Fast Pyrolysis
November 16-21, 2014
Atlanta, GA

Empirical approaches to predict fast-pyrolysis products by feedstock characterization and py-MBMS analysis

Hoyong Kim[1*], Carlos Aizpurua[1], Xueyong Ren [1, 2], Stephen Kelley[1], Hasan Jameel[1], Sunkyu Park[1]

[1] Department of Forest Biomaterials, North Carolina State University, Raleigh, NC 27695, USA
[2] College of Materials Science and Technology, Beijing Forestry University, Beijing, 100083, PR China.

* Presenting author: hkim36@ncsu.edu

Current fast pyrolysis simulation models are insensitive to coming feedstock and problematic to handle different types of biomass, even though it is generally agreed that chemical characteristics of incoming feedstock affected to the mass balance and the composition of fast pyrolysis products. Create a new model that sensitive to incoming feedstock is important because it will help us both finding a suitable feedstock and handling an unexpected biomass sources. Thus we try to develop a model that driven by chemical composition of incoming feedstock. In this study, the mass balance of pyrolysis products was correlated with incoming feedstock properties such as total carbon content (from 44.5% to 49.9%) and total ash content (from 0.3% to 7.4%) for ten biomass samples. Based on the product analysis after fluidized-bed fast pyrolysis at 500°C, char (from 6.3% to 24.6%) and water (from 8.2% to 26.4%) contents were significantly increased, while carbon and ash contents of feedstock are increased. On the other hand, organic oil contents were remarkably declined (from 58.0% to 29.2%). The chemical composition of the bio-oil was estimated based on the analysis of the pyrolysis molecular beam spectrometry (py-MBMS) at the same temperature of fast pyrolysis (500°C). For comparison, literatures dealt with similar reactor type were gathered and analyzed. Mineralized and demineralized feedstock will be used to reveal the role of the ash during fast pyrolysis, and we will explain the detail empirical models with process simulation results.

384565 Levulinate and Formate Salt Reactions during Thermal Deoxygenation (TDO)

Abstract:

Thermal deoxygenation (TDO) refers to a biochemical-thermochemical reaction scheme for the removal of oxygen from biomass carbohydrates. First, the carbohydrate is pre-treated via acid hydrolysis to yield levulinic and formic acids. These acids are then neutralized using a base, such as calcium or magnesium hydroxide, to form a salt. Slow heating of the salt mixtures to 450°C yields chars, carbonates, condensable products (crude bio-oil and water), and non-condensable gases. The crude TDO-oil has physical properties similar to a light petroleum crude which phase separates from water and has low total acid number (<2 mg KOH/g) and oxygen content (< 4 wt.%).

In this presentation, the conversion of solid-phase species is followed during heating in a tube furnace under inert atmosphere over the temperature range of 100-700°C. H^1 and C^{13} Nuclear Magnetic Resonance (NMR) and Diffuse Reflectance Infrared Fourier Transform Spectroscopy (DRIFTS) is used to monitor the solid-phase chemical composition of levulinate/formate salt mixtures. GC-MS is used to both monitor levulinate and formate conversion and identify volatile compounds captured during salt decomposition. TDO reactions are found to begin at temperatures above 200°C and further reactions initiate above 375°C. These reactions predominately yield dehydration and decarboxylation products, respectively. TDO reactions above 400°C begin to show a complex mixture of reaction products.

In-Situ Catalytic Upgrading of Pyrolysis Vapors in a Bench-Scale Fluidized Bed Reactor

Ravishankar Mahadevan, Rajdeep Shakya, Sneha Neupane and Sushil Adhikari

Biosystems Engineering Department, Auburn University, Auburn, Alabama

The conversion of biomass to pyrolysis oil (*i.e.*, bio-oil) and its subsequent conversion to gasoline and diesel range compounds have a significant potential in reducing the United States' dependence on imported petroleum products. However, the primary oil obtained from pyrolysis of biomass is acidic in nature, has a high oxygen content and low heating value, which requires an efficient upgrading technique to make it more attractive to be used as a fuel. In-situ catalytic upgrading is a promising technique to improve the quality of bio-oil, which offers an important advantage of direct conversion of biomass to liquid fuel in a single step, eliminating the need for multiple reactors to upgrade the bio-oil. In-situ catalytic upgrading combines pyrolysis and catalytic cracking in the same reactor, where the catalyst also serves as a heat carrier. In this study, catalytic pyrolysis experiments were performed in a bench scale fluidized bed reactor using MgO, CaO and ZSM-5 as in-situ catalysts. These catalysts displayed different effects on the pyrolytic product distribution, and their performance was compared based on the quality of bio-oil obtained in terms of its chemical composition, heating value, viscosity, pH, TAN, oxygen and water content. The stability of these bio-oils during storage was tested by subjecting them to accelerated aging conditions. While the anhydrosugars were reduced by all the catalysts to different extents, CaO significantly altered the quality of bio-oil by reducing the undesirable organic acid content. In comparison, MgO was a mild catalyst as it altered the product distribution slightly while ZSM-5 had no effect on the acid content in bio-oil although it produced a bio-oil with the greatest heating value at a significantly lower yield and higher water content. Accelerated aging test was performed as a tool to compare the efficacy of these in-situ catalysts on improving the stability of bio-oil, and it was observed that the bio-oil produced using CaO was the most stable and the aging index was significantly lesser when compared to the other catalytic bio-oils tested in this study.

IMPACT OF CELLULOSE ALLOMORPH ON ENZYMATIC SACCARIFICATION

Ting Cui[1, 2]*, Jihong Li*[1, 2]*, Zhipei Yan*[1, 2]*, Shizhong Li*[1, 2*]*, Menghui Yu*[1,2]

[1] *Institute of Nuclear and New Energy Technology, Tsinghua University, Beijing 100084, China*

[2] *Beijing Engineering Research Centre of Biofuels, Tsinghua University, Beijing 100084, China*

Abstract

Background

Bioconversion of cellulose into glucose or chemicals attracted extensive attentions for the sustainable development of human society in recent decades. The crystal allomorph of cellulose is a key factor to affect cellulose saccharification. However, due to the influence of lignin, hemicellulose and different characterizing methods of literatures, the effect of cellulose allomorph on cellulose saccharification is still confused. Thus, a systematic research on the effect of different cellulose allomorph on enzymatic saccharification is required.

Results

Multiple approaches, including ionic liquid (IL), ethylenediamine (EDA), glycerol and sodium hydroxide, were used to pretreat α-cellulose in this work. The properties of obtained cellulose (crystallinity index, lattice spacing, specific surface area and wettability) were characterized by X-ray diffraction, BET specific surface area analysis and water contact angle analysis, respectively. The distance of lattice spacing of cellulose III was larger than that of other cellulose samples. The crystallinity and water contact angles of all the cellulose samples were both in the following order: cellulose treated by IL < cellulose treated by NaOH < cellulose treated by EDA <α-cellulose < cellulose treated by glycerol. Cellulose treated by IL with a crystallinity index value of 20%, was very close to amorphous cellulose. After 72 h hydrolysis, the cellulose conversion rates ranged from 43% to 99%. Cellulose treated by IL exhibited a notably efficient hydrolysis profile, followed by cellulose treated by EDA.

Conclusion

Four cellulose allomorphs were prepared to investigate the effect of crystal form on the cellulose enzyme digestibility in this study. With the integrated experimental characterization, the most feasible enzymatic hydrolysis allomorph was cellulose III, followed by cellulose II, cellulose $I_α$ and $I_β$. When the crystal type of cellulose was the same, the crystallinity index was the dominant factor. Low crystallinity index enhanced the enzymatic hydrolysis efficiency of cellulose. Water contact angle was also a feasible characterization to evaluate the enzymatic hydrolysis efficiency of cellulose except for cellulose III. High wettability of cellulose enhanced the enzymatic hydrolysis when crystal allomorph of cellulosic samples was the same.

364398 Bio-Inspired Magnetic Pollen Particles Tailored with Adherents for Tuning Adhesion Strength and Optical Properties

Abstract:

Nature provides remarkable examples of bio-microscale particles with well-optimized structures and/or chemistries for the design of novel functional materials with effective adhesion to a variety of porous or dense surfaces. Herein, we reported the development of bio-inspired magnetic pollen grains as novel multifunctional microparticles with tailored specificity in both adherence and optical activity. The adhesion strength of as-synthesized magnetic pollen grain is significantly enhanced through the synergic interactions of the short-range van der Waals (VDW) force//hydrogen bonding, the long-range magnetic force and capillary force into one particle. The long-range magnetic force is dependent on the amount of Fe_3O_4 particles loading in the core, while the VDW force and the capillary interaction are tuned through coating polystyrene (PS) nanoparticle, chitin nanofiber and a layer of viscous pollenkitt on the shell surface, respectively. For optical activity, the as-synthesized magnetic pollen grains exhibit varied light-reflective intensities with different surface modifications. The difference of reflective indices of surface coating materials determined the pollen grain light-reflective properties, which were enhanced with pollenkitt coating, and dramatically reduced with PS nano-particle coatings. This present strategy provides us with new insight on the exploring of the generation of multifunctional particles with significantly-enhanced taggants capabilities, and it could further extend to other pollen grains and their biomimetic-replica high-aspect-ratio particles with numerous different nanoscale surface features for adhesion purpose.

388676 Nanocomposites from Modified Lignin

Abstract:

Lignin comprises 25~30% of biomass, which is the second largest among all the materials extracted from wood. However, due to some limitations, such as amorphous formation, structure variation and low solubility both in water and organic solvents, most lignin are burnt for heat. To use lignin as high-valued chemicals and materials, lignin needs to be modified. In this research, lignin modification using chemical such as butyric anhydride and 2-Bromoisobutyryl bromide to change its solubility in different organic monomers were conducted. Modified lignin is able to dissolve in styrene and methyl methacrylate monomer and therefore, can participate in the polymerization of polymers. The mechanism of this reaction is the reaction between the –OH group in lignin and anhydride in butyric anhydride or bromide in 2-Bromoisobutyryl bromide. More works can be done based on the water-soluble lignin. In this research, 1HNMR and FTIR are used as characterization methods to justify the change of lignin structure. 31PNMR is used to determine content of –OH group. What's more, Zetasizer is used to measure the particle size of PMMA using modified lignin as additive and SEM is used to observe its morphology. Results obtained from this research can be used to develop more expanded area of lignin application.

Synthesis of pH-sensitive Core-shell Latex with Biobased Dendrimer via Miniemulsion Polymerization

Suguna Jairam [1], **Zhaohui Tong*[2],

[1, 2] Agricultural and Biological Engineering Department, University of Florida, Rogers hall, Museum Road, PO Box 110570 Gainesville, FL. 32611

*ztong@ufl.edu

Abstract: The pH-responsible nanostructures have gained vast interest in the areas of gene therapy, drug delivery, and sensitive coatings. In this study, we synthesized pH-responsive core shell latex with the biobased dendrimer encapsulated dye via miniemulsion polymerization. Glycerol, main component in the biodiesel waste stream, was used to synthesize a novel triacrylate monomer, which was crosslinked with ethylene diamine to produce dendrimeric structures. Next, the dendrimers were copolymerized with styrene in the presence of Solvent Blue 59 dye via miniemulsion polymerization to produce pH-sensitive core shell latex with dye encapsulation. Nuclear Magnetic Resonance (NMR) analysis of glycerol triacrylate, dendrimer and polymers (at varying pH) showed the successful synthesis of pure monomer from glycerol, completed reaction with dendrimer, and pH sensitivity in polymer; respectively. Transmission Electron Microscopy (TEM) images in combination with zetasizer analysis data indicate uniform size distribution of polymer particles, successful dye encapsulation and release upon pH variation.

388838 Effects of Feedstock, Collection Method and Post-Condensation Filtration on the Stability of Pyrolysis Oil

Abstract:

The physical and chemical properties of the pyrolysis oil have been shown to be dependent on the feedstock composition. Forced aging tests were performed to understand the effects of feedstock, collection method and post-condensation filtration on the stability of pyrolysis oil. In this study, properties of pyrolysis oil produced using different ratios of pine clear wood and pine bark were measured and compared. The pyrolysis oil samples used for the study were produced using an auger reactor with multiple condensers. For each feedstock, three different fractions of the pyrolysis oil were collected by collecting from one or more of the condensers: total, fraction 1, and fraction 2.To understand the effects of post-condensation filtration both lab-scale and pilot scale centrifuge were used. The Pilot scale centrifuge can be used as a three-phase separator [Light liquid-Heavy liquid-Solids] or a two-phase clarifier [Liquid-Solid]. The pyrolysis oil is a micro emulsion of oil and water which makes it difficult to separate water from the oil. So Emulsion destabilization studies were performed on the lab-scale centrifuge. The effect of feed temperature on the separation efficiency of the pilot-plant centrifuge was also studied. Samples were characterized to determine the physical properties such as water content, pH, density, viscosity, and solids content. For chemical characterization, FTIR spectroscopy was utilized to identify the functional groups present in the samples. Optical microscopy was used to determine the particle size distributions of solids in these samples.

378362 The Effect of Hydrogen-Donor Agent in Catalytic Pyrolysis of Lignin to Hydrocarbons

Abstract:

The effect of hydrogen-donor agent in catalytic pyrolysis of lignin to hydrocarbons

Xianglan Bai[1], Shuai Zhou[2], Robert C. Brown[1]

[1]Department of Mechanical Engineering, Iowa State University, Ames, IA 50011, USA

[2]Bioeconomy Institute, Iowa State University, Ames, IA 50011 USA

Catalytic pyrolysis of lignin using zeolite as the catalyst is known to convert lignin into aromatic hydrocarbons. However the yield of the aromatic is usually low and deactivation of catalyst is problematic. Our previous study showed that reactive phenols derived from lignin during fast pyrolysis subsequently repolymerize in the vapor phase. Therefore it is important to stabilize the reactive phenols prior to following catalytic upgrading. In the present study, lignin is co-pyrolyzed with tetralin in a tandem micropyrolyzer. Tetralin is capable of releasing hydrogen atoms at atmosphere pressure, thus high pressure is not required. The pyrolysis vapor is then sent to a downstream catalytic bed that is filled with HY or HZSM-5 zeolite. It is found that the presence of the hydrogen-donor agent promotes saturation of vinylphenol, demethoxylation and demethylation of phenols. Catalytic upgrading of the stabilized pyrolysis vapor improves hydrocarbon yield from lignin and inhibites coke formation on catalyst. Saturation of vinyl groups and demethoxylation by hydrogen-donors contributes to the observed changes.

Effect of Operating Parameters on the Fast Pyrolysis of Pakistani Maize Stalk in Bubbling Fluidized Bed Reactor

Khurram Shahzad[1*], Najaf Ali[2], Mahmood Saleem[3], Arshad Chughtai[4]

1. Centre for Coal Technology, University of the Punjab, Pakistan
2. NFC Institute of Engineering & Technology, Pakistan
3. Institute of Chemical Engineering & Technology, University of the Punjab, Pakistan
4. School of Chemical and Materials Engineering, NUST, Pakistan

ABSTRACT

Fast pyrolysis is one of the prospective techniques employed to get energy and valuable chemicals from biomass. The yield and composition of pyrolysis products depend on the characteristics of feedstock and process operating parameters. A lab scale pyrolysis system consists of bubbling fluidized bed reactor (capacity 2kg/hr) is being designed. Effect of particle size and operating temperature on the fast pyrolysis of Pakistani maize stalk has been investigated. Experimentation was done with the temperature range of 360-540°C and particle size of 1-2 mm (SMD). Both parameters are found to have significant effect on the yield and characteristics of bio oil produced under the bubbling fluidized bed conditions. Bio oil yield increased with the increase of temperature followed by a decreasing trend. Maximum yield (41 wt. %) of bio oil was obtained at 540°C with particle size of around 1.0 mm. High temperatures resulted in high ratios of char and gas. GC-MS analysis of bio oil identified the large number of compounds including butanal, 2,3-Butanedione, Benzene, 1-Butanol, Acetic Acid, 3-Pentanone, Acetol, Toluene, 2-Butenal, 2-methyl-methyl pyruvate, Propanoic acid, Ethylbenzene, p-Xylene, Cyclopentanone, 2-methyl- o-xylene, 2-Butanone, 4-hydroxy-Furfural, 2-Furanmethanol, Benzene, 1,2,3-trimethyl-1-Heptanol, Benzaldehyde, 4-methyl-phenol, guaiacol, o-cresol, Naphthalene, Phenol, 2-methoxy-4-methyl-phenol, 2,4-dimethyl-phenol, 4-ethyl-naphthalene, 2-methyl-2-Furancarboxaldehyde, 5-(hydroxymethyl)-syringol, Vanillin, Ethanone,1-(4-hydroxy-3-methoxyphenyl), 3',5'-Dimethoxyacetophenone, 2-Propanone, 1-(4-hydroxy-3-methoxyphenyl), 4-Hydroxy-2-methylacetophenone, 4-Hydroxy-3-methylacetophenone, Levoglucosan, syringaldehyde, Ethanone, 1-(4-hydroxy-3,5-dimethoxyphenyl). Gas Chromatography analysis of non-condensable gases produced from the fast pyrolysis process is also being done and discussed.

1. INTRODUCTION

The massive supplies of biomass available in Pakistan have the potential to displace the substantial amounts of fossil fuels. Biomass is an indigenous source of energy which may be available on renewable basis by natural processes and human activities such as organic wastes. In the recent years, special consideration has been given to the change of biomass into bio fuels. Biofuel is renewable and does not add carbon dioxide to the atmosphere as a result to the greenhouse effect. Bio energy can be produced from various types of biomasses including forest residues, agricultural wastes, and aquatic plants. Biomass has been recognized as 4th major energy source in the world and contributes around 14 % of primary energy consumption in the world whereas its contribution in the developing countries ranges from 40-50 % but in the form of traditional fuels, it delivers over 90 % of total energy in the developing countries, for example fuel wood and dung [1-3]. In order to compensate the reduction of fossil fuels and meet the national energy requirements in these countries, development of alternate energy sources is important [4].

Pyrolysis is a thermochemical conversion process which is used to convert biomass into liquid bio fuel. It is thermal decomposition of biomass by heat in the presence of inert atmosphere to yield liquid, solid and gaseous products. It is not possible to define pyrolysis specifically because in early times it named as carbonization in which primary product is charcoal. Today the term pyrolysis is understood for the process in which liquid fuel is the key product. In present energy development, pyrolysis process has attained distinct consideration because of biomass conversion into energy rich products. Pyrolysis is thought to be a promising route for conversion of biomass into bio fuels and chemicals [1]. The basic reaction taking place during pyrolysis process is:

$$\text{Biomass} \longrightarrow \text{Char} + \text{Volatiles}$$

The process of pyrolysis may be categorized as fast pyrolysis, conventional pyrolysis and flash pyrolysis. Conventional pyrolysis may be known as slow pyrolysis or carbonization which takes place at reaction conditions of low temperature, slow heating rate and high vapor residence time. Charcoal is known as the principal product of this pyrolysis process. Fast pyrolysis process is conducted to yield liquid fuel at moderate reaction temperature and short vapor residence time whereas flash pyrolysis follows at high heating rate, very short vapor residence time of few seconds or less and very fine feed particle size (≤ 2 mm) [2-4].

2. EXPERIMENTAL

2.1 MATERIALS AND METHODS

Maize stalk used as feedstock is collected from Pakpattan, District Sahiwal, Pakistan. and shredded into coarse particles. The samples are finally ground in the grinding mill upto particle size of <2 mm and are air dried having moisture content of <10%. Silica sand is used as bed material in the fluidized bed reactor to perform experiments and size of sand is -40, +60 mesh and density 2600 kg/m^3. Nitrogen is used as fluidizing gas in the silica sand bed.

Ultimate and proximate analysis of the maize stalk is carried out at GC University, Lahore and NFC IET, Multan using Vario Micro CHNS Serial No.15071001 analyzer and Leco TGA701 analyzer. The higher heating value (HHV) is determined from Leco AC 500 auto bomb calorimeter. The results of CHNS, proximate analysis and HHV are explained in the table 1.

Table 1: Ultimate and Proximate Analysis of Maize Stalk

Ultimate Analysis (wt % as received)		Proximate Analysis (wt% as received)	
C	40.55	VM	63.63
H	5.64	FC	30.34
O	48.33	Ash	3.88
N	0.9	Moisture	0.68
S	0.01	HHV(MJ/Kg)	15.99

Chemical composition of bio oil is determined using GC-MS analyser and Fourier Transform Spectroscopy (FTIR) is applied to identify the functional groups and bonds in the bio oil liquid. Karl Fischer Titrator is used to determine the water contents in the bio oil using hydranal as titrant. Composition of incondensable gas obtained from the pyrolysis of maize stalk is found using GC analyser.

2.2 FAST PYROLYSIS SETUP

Fast pyrolysis bench scale plant (fig.1) with feed capacity of 2 kg/h is designed, developed and operated at the Institute of Chemical Engineering & Technology, University of the Punjab, Lahore Pakistan. The major components of the system are biomass feeding system, fluidized bed reactor, heating system, gas solid separation system, pyrolysis vapors condensation system and instrumentation and control system. In the fast pyrolysis process, screw feeder is used to transport the regulated biomass feedstock into the fluidized bed reactor and preheated inert nitrogen gas at about 350-400 °C is passed through the bottom of the reactor as a fluidization gas for the sand bed. Pyrolysed volatiles along with charcoal leave the reactor from the top. The solid char particles are separated from the gas in the cyclone separators and pyrolysis gas is first condensed in the water cooled condenser and then in dry ice condenser to obtain pyrolysis liquid. The non-condensable gas is passed through a filter to remove fine solids and then samples of gas are taken after every 15 min for analysis. The remaining gas is either burnt in the flare or vented out. The fluidized bed reactor is made of stainless steel SS 316 L and has diameter of 100 mm and height of 750 mm. The reactor is heated indirectly with carbide heating rods to maintain the pyrolysis reaction temperature.

The feeding system is comprised of a hopper with nominal capacity of 0.03 m^3, stirrer to entrain the biomass and screw feeder having the variable speed to drive the feedstock into the reactor. The biomass is dried at 105°C for overnight prior to run in the pyrolysis experiments. There should be positive pressure between hopper and fluidized bed reactor to avoid the back flow of pyrolysis vapors into the feeding hopper. The distributor plate is made of stainless steel with pore size of about 100 μm to retain the silica sand bed. Nitrogen gas is heated into the preheater at 350-400°C to establish the temperature in the reactor. The biomass is directly fed to the bed in the reactor after achieving the designated temperature. The temperature at the cyclones is maintained at about 400°C to avoid the condensation of pyrolysis vapors in order to minimize the secondary cracking reactions responsible for producing surplus gas.

Figure 1: Experimental Setup

The maize stalk is pyrolysed in the fluidized bed reactor at temperature 360-540 °C and yield of bio oil, charcoal and uncondensable gas are determined from condensed liquid from water and dry ice condensers, from two cyclones and by difference. These calculations are based on mass balance from the different components of pyrolysis system.

3. RESULTS AND DISCUSSION

3.1 Effect of Operating Parameters on the Yield of Pyrolysis Products

3.1.1 Effect of Pyrolysis Temperature

Fast Pyrolysis of maize stalk is carried out in fluidized bed pyrolysis reactor. Bio oil, charcoal and gas is collected after each experiment and yield of individual product calculated from the mass balance. Figure 2 explains the effect of temperature on pyrolysis product distribution of maize stalk in the same pyrolysis system as mentioned above. As presented in fig. 2, rise of temperature from 360 to 540°C, yield of char decreases while yield of gas increases. The increase in the reaction temperature from 360 to 490°C, the yield of pyrolysis oil rises, even though at the considerable greater temperature of 540°C, there is a drop in yield of liquid oil. The value of pyrolysis liquid obtained is maximum (41 wt.%) at a pyrolysis temperature of 490°C.

These trends of pyrolysis liquid, char and non- condensable gas yields with temperature are similar to the results in the literature [5-9].

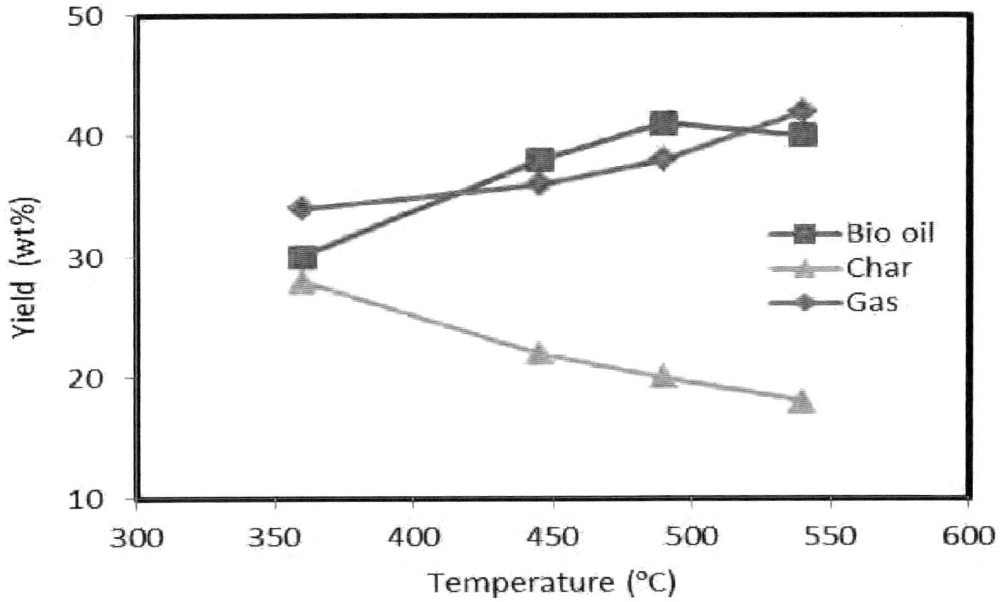

Figure 2: Effect of temperature on maize stalk

The bio oil from the fast pyrolysis of maize stalk is in the homogenous phase and dark brown in color. FTIR absorbance spectra peaks of pyrolysis liquid obtained is shown in the fig.3.

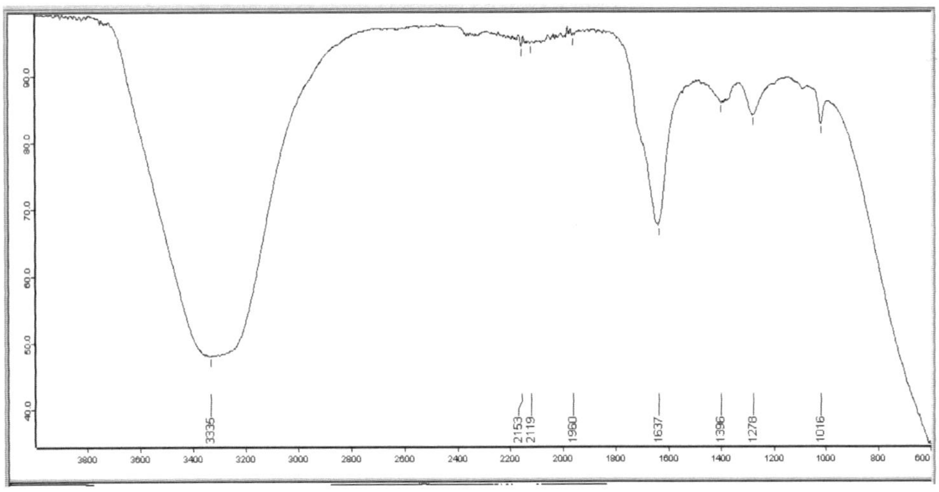

Figure 3: FTIR Absorbance Spectra of Bio oil of Maize Stalk

Table 2: Functional Groups from FTIR Spectra of Bio Oil from Maize Stalk

Wavenumber (cm^{-1}) actual	Wavenumber (cm^{-1}) Range	Functional Group	Bond
3333 (s, b)	3500-3200	Phenols, alcohols	O-H
3333 (m, n)	3300-2500	Carboxylic acids	O-H
1637(s, n)	1750-1625	Ketones	C=O
1637(m, n)	1750-1625	Aldehydes	C=O
1387 (m, n)	1450-1470	Alkanes	C-C
1277(m, n)	1350-1370	Alkanes	C-C
1016(s, s)	1000-1320	Alcohols, carboxylic acids, ethers, esters	C-O

In this FTIR Spectra, absorbance vs wavenumber peaks obtained are analysed to identify the functional groups or bonds present in the pyrolysis liquid and are shown in the table 2. The compounds found in the bio oil are phenols, alcohols, carboxylic acids, ketones, esters and ethers.

3.1.2 Effect of Particle Size

Pyrolysis of biomass for feed sizes of 1.0, 1.5 and 2.0 mm is depicted in the fig.4 for pyrolysis products distribution. It is evident that bio oil decreases with particle size and also visible that effect is linear over the selected range of particle sizes.

It is shown from the graphs that the bio oil yield is decreased by increasing the feed size of maize stalk whereas char and gas yield are increased. For maize stalk, the bio oil yield is decreased from 42 to 35 wt.%, the char yield from 22 to 25 wt.% is increased and gas yield increased from 36 to 40 wt.% by the increase of particle size of feedstock. This effect of particle size of biomass is conceived due to the heating rate. By increasing the size of feed particle, distance from the surface of particle to its center increases which impedes the rapid heat transfer indicating the incomplete pyrolysis reaction there and hence the pyrolysis volatiles are cut down which decreases the pyrolysis oil yield.

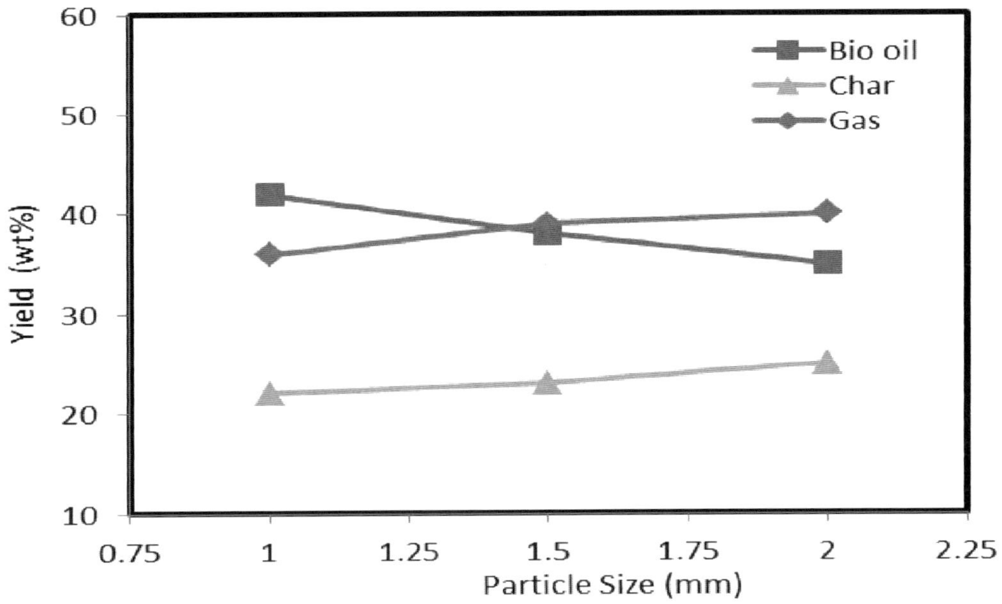

Figure 4: Effect of particle size on maize stalk

4. CONCLUSIONS

Fast pyrolysis of maize stalk is investigated in locally developed fluidized bed reactor and effect of operating parameters such as pyrolysis temperature and particle size are studied. It is observed from the experiments that reaction temperature has significant influence on the distribution of pyrolysis products. The yield of pyrolysis liquid increases with the temperature to maximum and then decreases while yield of char decreases and gas increases. In addition to pyrolysis temperature, feed particle size also affects the yields of pyrolysis process. The pyrolysis liquid yield decreases by the increase of particle size and yields of bio char and gas increases. The maximum yield of pyrolysis liquid is obtained at the optimum pyrolysis temperature of 500°C and feed size of 1mm.

REFERENCES

1. Goyal HB, S.R., Seal D. communicated to Haworth Press, USA, 2006, *Thermochemical conversion of biomass to liquids and gas*. 2006: Monograph.
2. Jayasinghe, P. and K. Hawboldt, *A review of bio-oils from waste biomass: Focus on fish processing waste.* Renewable and Sustainable Energy Reviews, 2012. **16**(1): p. 798-821.
3. Mohan, D., C.U. Pittman, and P.H. Steele, *Pyrolysis of Wood/Biomass for Bio-oil: A Critical Review.* Energy & Fuels, 2006. **20**(3): p. 848-889.
4. Bridgewater, A.V., *Biomass fast pyrolysis.* Thermal Science, 2004. **8**(2): p. 21-50.
5. Zheng, J.-L., *Pyrolysis oil from fast pyrolysis of maize stalk.* Journal of Analytical and Applied Pyrolysis, 2008. **83**(2): p. 205-212.
6. Xueyuan, B., et al., *Fast pyrolysis of corn stalk for bio-oil in a plasma heated fluidized bed.* Transactions of The Chinese Society of Agricultural Engineering, 2005. **21**(12): p. 127-130.
7. Park, H.J., Y.-K. Park, and J.S. Kim, *Influence of reaction conditions and the char separation system on the production of bio-oil from radiata pine sawdust by fast pyrolysis.* Fuel Processing Technology, 2008. **89**(8): p. 797-802.
8. Wang, X., et al., *Biomass pyrolysis in a fluidized bed reactor. Part 2: Experimental validation of model results.* Industrial & engineering chemistry research, 2005. **44**(23): p. 8786-8795.
9. Garcia-Perez, M., et al., *Characterization of bio-oils in chemical families.* Biomass and Bioenergy, 2007. **31**(4): p. 222-242.

Characterization of Bio-oil and Associated Aqueous Phase Derived from Switchgrass Pyrolysis

Shoujie Ren[1], Philip Ye[1], Abhijeet P. Borole[2], Pyoungchung Kim[3], Niki Labbé[3]
[1] Biosystems Engineering and Soil Science, The University of Tennessee Knoxville, TN, 37996
[2] Bioscience Division, Oak Ridge National Laboratory, Oak Ridge, TN 37831
[3] Center for Renewable Carbon, University of Tennessee, Knoxville, TN 37996

Abstract: Bio-oil is carbon based liquid product from biomass pyrolysis. Bio-oil is considered to be a promising substitute of crude oil in the applications as source of chemicals and as fuels. However, bio-oil has high water content, high viscosity, low heating value, low pH value, and high oxygen content, leading to poor and inconsistent physicochemical properties. Bio-oil consists of hundreds of chemicals with wide range of molecular weight. Bio-oil aqueous phase accounts for about 60% of crude bio-oil and has higher oxygen content. Conventional analytic method of GC-FID or HPLC only can quantify small part of chemicals of bio-oil aqueous phase as it contains volatile and non-volatiles compounds. The lack of adequate norms and standards for bio-oil and the bio-oil aqueous phase analysis have inhibited their industrialized production and application. In this study, the physical properties of bio-oil and aqueous phase from swichgrass pyrolysis were investigated and a combination technique of organic solvent extraction, GC/MS, GC-FID and HPLC-PDA was used to identify and quantify chemicals in bio-oil aqueous phase in order to comprehensively understand the chemical distribution in bio-oil aqueous phase. Two organic solvents, ethyl acetate and chloroform were investigated as solvent to extract chemicals from aqueous phase and extracted chemicals were identified by GC/MS. The main chemicals in aqueous phase were classified into 8 categories including acids, aldehydes, ketones, furans, phenols, ethers, sugars, and alcohols. Seventeen major compounds from different categories in bio-oil aqueous phase were quantified by GC-FID and HPLC-PDA using external standard chemicals.

Keywords: Bio-oil aqueous phase, HPLC-PDA, GC/MS, GC-FID

Introduction

Current fuel supply is greatly depends on fossil fuel. However, the limited amount of fossil oil reservation and greenhouse gas emission in the combustion of fossil oils motivate the scientists and researchers to look for alternative and clean energy sources. Biomass is one of important and large amount renewable fuel sources and is considered carbon neutral as its carbon is recycled from the atmosphere [1]. Pyrolysis which is thermal chemical technology generally conducted at temperatures ranged from 400 to 600 °C in the absence of oxygen and produced bio-oil, biochar, and syngas is one of technologies to conversion biomass to useful energies or chemical products [2].

Bio-oil produced from biomass pyrolysis is carbon based liquid product. Bio-oil is considered to be a promising substitute of crude oil in the applications as source of chemicals and as fuels. However, bio-oil has poor physical-chemical properties such as high water content and viscosity, low heating value and pH value, and high oxygen content that limited its application [3]. The bio-oil is a mixture of complex chemical components that hundreds of chemicals have been identified. The bio-oil can be separated to organic phase and aqueous phase by adding water or organic solvents [4-7]. Most polar compounds such as carboxylic acids and anhydrosugars are extracted to aqueous phase. Other compounds such as phenolic, furans, alcohols and ketones in aqueous phase were also identified even after organic solvent extraction reported by published paper [4]. Due to the high water content and low concentration of compounds existed, bio-oil aqueous phase is low value compared with crude oil and organic phase and cannot directly as fuel. To improve the economics of biomass pyrolysis, the application of bio-oil aqueous phase for hydrogen or value chemicals production has drawn interest in last several years [8-12]. However, the comprehensive analysis for the bio-oil aqueous phase is lack. So for better understanding the application of bio-oil aqueous phase, comprehensive characterization of bio-oil aqueous phase is very important.

Due to the complex of components and volatile and non-volatile properties of chemicals in bio-oil, the GC/MS, GC and HPLC are generally use to identify and quantify the components in the bio-oil. Bio-oil aqueous phase contains large amount of water and small fraction of chemical compounds generally less than 10wt% which is not suitable for the analysis by GC/MS directly. In this study for better knowing chemicals existed in the bio-oil aqueous phase we used organic solvents to extract the chemicals of aqueous phase and the extracted samples were analyzed by GC/MS for identifying the compounds. The major compounds were further quantified by using HPLC-PDA and GC-FID. The physical properties of bio-oil and bio-oil aqueous phase were also investigated.

Materials and methods

2.1 Materials

Air-dried switchgrass (Panicum virgatum L.) obtained from a local producer in eastern Tennessee was used for the bio-oil production. The water content of switchgrass was 7-8 wt%. Before pyrolysis switchgrass was ground to less than 2mm particles size. The details of compositions of switchgrass can be found in the previous report [13].

Ethyl acetate and chloroform used as organic solvents for chemicals extraction were purchase from Thermo Fisher Scientific (Waltham, MA). External standards used for quantifying compounds were purchased from Sigma-Aldrich (St. Louis, MO).

2.2 Bio-oil production and aqueous phase separation

A semi pilot-scaled auger pyrolysis system (Proton Power, Inc., Lenoir City, TN) equipped with a feeding system, a rectangular auger reactor, a pyrolysis vapor condensation section, and a biochar collector was used to switchgrass pyrolysis for bio-oil production. The switchgrass pyrolysis was conducted at 500 °C with a residence time of 90 s and feeding rate of 10 kg h^{-1} in the presence of N_2.

The crude bio-oil was mixed with DI water in 1:4 weight ratio for bio-oil aqueous phase separation. The mixture was shaken vigorously and placed at 4°C overnight. Then the mixture was centrifuged in an IEC Clinical centrifuge (model 120) at 5000 rpm for 30 minutes to ensure the phase separation.

2.3 Physical properties analysis of crude bio-oil and aqueous phase analysis

The crude bio-oil and aqueous phase properties of density, pH, viscosity, water content, solid content, ash content, and total acid number (TAN) were measured in triplicate. Density was measured according to the standard method of ASTM D1217, and pH was measured with an Extech pH meter. A Schott TitroLine Karl Fischer volumetric titrator was used to measure water content according to ASTM D4377 standard. Viscosity was measured at 40°C with serialized Schott Ubbelohde capillary viscometers according to ASTM D445 standard. Ash content was measured according to ASTM D482 standard at 575°C. The solid content was determined according to Boucher et al. [14]. TAN was measured by titrating bio-oil in solvent of isopropyl alcohol and toluene with 0.1M KOH isopropyl alcohol solution to an end point of pH 11 according to ASTM D664 standard.

2.4 Organics extraction from bio-oil aqueous phase

Two organic solvents, ethyl acetate and chloroform, were first mixed with bio-oil aqueous phase in the volume ratio of 1:1 and 2:1 in separating funnel and shaken vigorously for completely mixing. Then the separating funnel was settled for 24 hours for phase separation. For calculating mass balance, the weights of organic solvents and bio-oil aqueous phase before and after extraction were also recorded.

2.5 GC/MS analysis for crude bio-oil and extracted chemicals from aqueous phase

The crude bio-oil and extracted chemicals from bio-oil aqueous phase was analyzed by GC/MS. A Shimadzu GC–MS (QP2010S) with a Restek Rtx-5MS capillary column was used. The column was programmed at 45°C for 3min, then at 5°C/min to 150°C without holding, then 10°C/min to 260°C and hold 7min. The inlet was set at 240 °C, and sample injection was made in split mode (1:20). The compounds were identified by comparing their mass spectra with those from the National Institute of Standards and Technology (NIST) mass spectral data library.

2.6 GC-FID and HPLC analysis for bio-oil aqueous phase

The HPLC system (Jasco 2000Plus, Jasco analytical instruments) equipped with PU-2089S Plus pump, a MD-2018 Plus Photodiode Array detector (PAD), a RI-2031 Plus intelligent RI detector, and an AS-2055 Plus auto sampler was used to analyze aqueous phase oil. The liquid chromatography was conducted at 50.0°C using Bio rad column HPX-87H (300 x 8 mm) column. The injected sample was 20μl. The mobile phase was 5mM H_2SO_4 with 15 % acetonitrile (v/v) in deionized water with a flow rate of 0.6mL/min.

GC-FID with a HP-5 column was used for quantifying volatile compounds of bio-oil aqueous phase. The same program with the GC/MS was used in GC-FID. Compounds were quantified using external standards in both HPLC and GC-FID analysis.

Results and discussion

3.1 Bio-oil production and bio-oil aqueous phase separation

The liquid fractions from pyrolysis were condensed from three condensers and collected. All collected fractions were combined to obtain the crude bio-oil. The biochar was collected and weighed after the pyrolysis. Non-condensable gases were burned out at the end of condensers during the pyrolysis. The weight of non-condensable gases was calculated based on the difference between the weight of pyrolyzed switchgrass and total amount of bio-oil and biochar. In this study, we obtained in 50 wt% of bio-oil, 29 wt% of char and 21 wt% of non-condensable gases using Auger pyrolysis of switchgrass.

The crude bio-oil was mixed with DI water in 1:4 weight ratio to separate into two phases: aqueous phase and organic phase. In the crude bio-oil water content was about 42.3 % and chemicals content about 55.7 %. After separation, about 35.1 % and 36.7 % of chemicals and water in crude bio-oil was extracted to aqueous phase. Figure 1 shows the fraction distributions before and after separation in the sample.

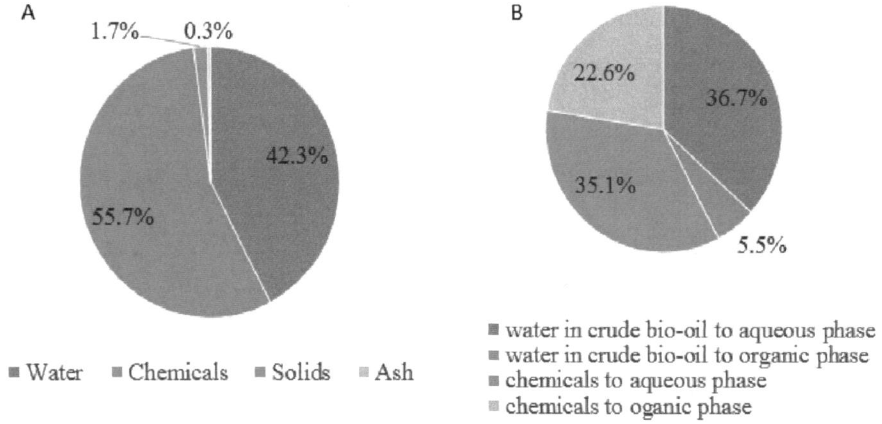

Figure 1 Fraction distribution before and after aqueous phase separation (wt%)

3.2 Physical properties analysis of crude and bio-oil aqueous phase

Table 1 shows the details of physical properties of crude and bio-oil aqueous phase. The water content in crude bio-oil is about 42 wt% which is higher than the bio-oil produced from fluidized bed pyrolysis. The pH value, density and ash content of crude bio-oil observed are about 2.8, 1.13g/ml, and 0.31 wt% respectively, which are similar with the previous reported results [15]. The viscosity of crude bio-oil is about 6.5 cSt at 40°C, lower than previous report. That is because crude bio-oil in this research contains high water. The observed total acid number (TAN) of crude bio-oil is about 137. Due to the water added in the aqueous phase separation, the water content and the pH value of aqueous phase in bio-oil aqueous phase is about 92 wt% and

3.02, much higher than crude bio-oil while the viscosity and TAN number of bio-oil aqueous phase are reduced comparing with crude bio-oil.

Table 1 Properties of crude and bio-oil aqueous phase

	Crude bio-oil	Bio-oil aqueous phase
Water content (wt%)	42.3±0.66	91.7±1.03
Total solid (wt%)	1.7±0.25	Not detected
pH value	2.8±0.07	3.0±0.01
Density (g/ml)	1.1±0.001	1.0±0.004
Ash (wt%)	0.3±0.04	0.1±0.004
Viscosity at 40 °C centistokes (cSt)	6.5±0.82	0.8±0.01
TAN, mg KOH/g	137.4±2.96	30.1±1.28

3.3 Chemical extraction of bio-oil aqueous phase by organic solvents

After the extraction by organic solvents, the solvents were evaporated by rotary evaporator and the organics extracted from aqueous phase oil were collected and weighted. The organic distribution in the organic solvent and water phase is shown in Figure 2. From the Figure 2, we can see that around 57 wt% and 70 wt% of organic compounds in aqueous phase oil was extracted by ethyl acetate at the ratio of 1:1 and 2:1 while only about 46 wt% and 54 wt% of organic compounds in aqueous phase oil was extracted by chloroform. It indicates that ethyl lactate has better extraction for chemicals extraction from bio-oil aqueous phase.

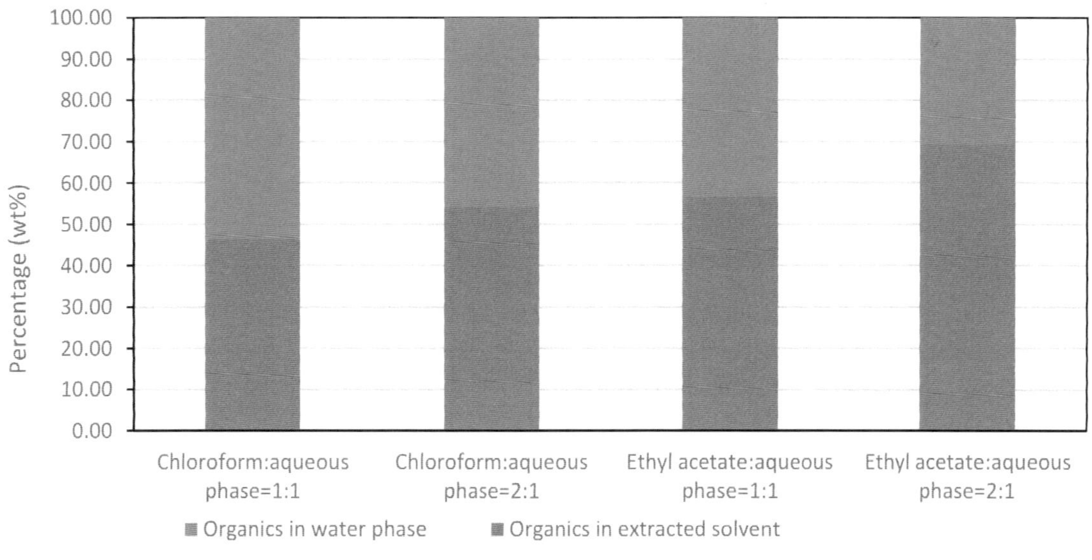

Figure 2 Organic distribution in solvents and water phase

3.4 GC/MS analysis for crude bio-oil and extracted organics from aqueous phase by organic solvents

Table 2 shows the identified compounds for crude bio-oil and extracted organics from aqueous phase by organic solvents. The identified compounds was categorized to nine groups, acids, alcohols, aldehydes, esters, furans, ketones, phenolics, sugars, and PAHs (poly aromatic hydrocarbons). The extracted organic from aqueous phase analyzed by GC/MS indicates that large amount of alcohol, furans, ketones, and phenolics were extracted by organic solvent. Acids, esters, aldehydes, and sugars were also detected in GC/MS but the relative area% is very low. Organic acids and levoglucosan were detected in crude bio-oil with significant amount and can be extracted to aqueous phase oil but in extracted organics for aqueous phase oil they were not detected or very low concentration, indicating that they were not extracted by ethyl acetate and chloroform but remained in the aqueous phase.

Table 2 Identified compounds in crude oil and extracted organics from aqueous phase by GC/MS

Categories	Compounds	Crude oil	Extracted by ethyl lactate	Extracted by chloroform
Acids	Acetic acid	X		
	propionic acid	X		
	Succinic acid, methyl-	X		
	vanilic acid	X	X	
	Benzoic acid, 3-hydroxy-4-methyl-	X		
Alcohols	1,3-Propanediol	X	X	
	1,3-Propanediol, 2-(hydroxymethyl)-2-methyl-	X		
	3-Hexanol, 2,4-dimethyl-	X		
	Cyclodecanol	X	X	X
	3-Cyclobutene-1,2-dione, 3,4-dihydroxy-	X	X	X
	1,2-Cyclopentanedione, 3-methyl-	X	X	X
	5-Isopropenyl-2-methylcyclohexanol	X		
	Cyclohexanol, 2,3-dimethyl-	X	X	X
	4-Cyclopentene-1,3-diol, trans-	X		
	3-Nonyn-1-ol	X		
	2-Hexen-1-ol, 2-ethyl-	X	X	X
	Cyclododecanol	X	X	
	Bicyclo [3.1.1] hept-3-en-2-ol, 4,6,6-trimethyl-	X	X	
	Cyclohexanol, 2-methyl-5-(1-methylethenyl)-	X	X	X
	1-Cyclohexene-1-methanol, 4-(1-methylethenyl)-	X		
Aldehyde	2-Furaldehyde, 5-methyl-	X	X	X
	5-Methyl-2-hexanone	X	X	
	1-Cyclohexene-1-acetaldehyde, 2,6,6-trimethyl-	X	X	X
Esters	Butyric acid, 3-methyl-, allyl ester	X		
	Benzoic acid, 4-hydroxy-3-methoxy-, methyl ester	X	X	X
	2-Octynoic acid, methyl ester	X	X	X
	p-Menth-8-en-2-ol, acetate	X		
	Pentanoic acid, 4-methyl-, ethyl ester	X		
Furans	Furfural	X	X	X
	2(5H)-Furanone	X	X	X
	2(5H)-Furanone, 5-methyl-	X	X	
	Benzofuran, 2,3-dihydro-	X	X	X
Ketones	1-Hydroxy-2-butanone	X	X	X
	2-Cyclopenten-1-one, 3-methyl-	X	X	X
	Cyclohexanone	X	X	X
	2-Butanone, 3,3-dimethyl-	X	X	X
	2-Cyclopenten-1-one, 3-ethyl-2-hydroxy-	X	X	X
	Cyclohexanone, 4-hydroxy-	X	X	
	2,5-Cyclohexadiene-1,4-dione, 2-methyl-5-(1-methylethyl)-	X		X
PAHs	Anthracene	X		
	Pyrene	X		
Phenolics	Phenol	X	X	X
	Phenol, 3-methyl-	X	X	
	Phenol, 2-methyl-	X	X	
	Phenol, 2-methoxy-	X	X	X
	Phenol, 3-ethyl-	X	X	X
	Phenol, 3-(1-methylethyl)-	X		
	1,2-Benzenediol	X	X	X
	Phenol, 2-methoxy-4-methyl-	X	X	
	Phenol, 2,3,5,6-tetramethyl-	X		
	Phenol, 4-ethyl-2-methoxy-	X	X	X
	2(1H)-Naphthalenone, octahydro-, trans-	X		
	Phenol, 2-methoxy-5-(1-propenyl)-, (E)-	X	X	X
	Phenol, 2,6-dimethoxy-	X	X	X
	Benzaldehyde, 3-hydroxy-4-methoxy-	X	X	X
	1,2,3-Trimethoxybenzene	X	X	X
	1,4-Benzenediol, 2-(1,1-dimethylethyl)-	X	X	X
	Benzene, 1,2,3-trimethoxy-5-methyl-	X	X	
	Phenol, 2,6-dimethoxy-4-(2-propenyl)-	X	X	X
	Benzaldehyde, 2,4,5-trimethoxy-	X	X	X
Sugars	d-Mannitol, 1,4-anhydro-	X	X	
	1,6-Anhydro-.beta.-D-glucopyranose (levoglucosan)	X		

3.5 Chemicals quantification by HPLC and GC-FID

According to the GC/MS analysis for crude and bio-oil aqueous phase analysis, total 17 compounds including 3 acids, 3 furans, 2 alcohols, 6 phenolics, 2 ketones and 1 anhydrosugars in bio-oil aqueous phase were quantified by GC-FID and HPLC-PDA. Among these compounds eight major compounds were quantified using external standards by HPLC-PDA and another 9 compounds were quantified by GC-FID. The results for quantification are shown in Table 3. Levoglucosan and acetic acid were detected large amount in bio-oil aqueous phase which have concentration of 15.33 and 11.96 g/L and account for about 18.47 ad 14.41% of organics in aqueous phase. Other quantified compounds were detected low concentration in bio-oil aqueous phase, generally less than 3 g/L. The total amount of quantified compounds by HPLC was about 37g/L based on bio-oil aqueous phase and contributed about 44.57% of organics in aqueous phase. 1,3-propanediol, 2(5H)-furanone, and 1-hydroxybutanone have relatively high concentrations comparing with other 6 identified compounds in bio-oil aqueous phase by GC-FID. Total amount of quantified compounds by GC-FID was about 6.02 g/L based on bio-oil aqueous phase and contributed about 7.26% of organics in aqueous phase. The concentration of quantified compounds by HPLC and GC-FID was about 43.01g/L of bio-oil aqueous phase and more than 51% organics in aqueous phase was quantified.

Table 3 Mass content of 17 major chemical compounds in bio-oil aqueous phase

Quantification method	Major chemicals	Concentration based on aqueous phase (g/L)	Percentage based on organics in aqueous phase (wt%)
HPLC	Furfural	1.01	1.22
	1,2-benzendiol	1.77	2.13
	Phenol	1.8	2.17
	Levoglucosan	15.33	18.47
	Acetic acid	11.96	14.41
	Proponic acid	1.89	2.28
	Vanillic acid	2.69	3.24
	HMF	0.54	0.65
	Total	36.99	44.57
GC	Phenol, 2-methoxy-	0.25	0.3
	2-methyl-4-methyphenol	0.07	0.08
	Cyclohexanone	0.07	0.08
	3-methyl-1,2-cyclophetandiol	0.46	0.55
	2,6-Dimethoxyphenol	0.26	0.31
	1,3-propanediol	1.84	2.22
	3-ethylphenol	0.56	0.67
	2(5H)-Furanone	1.17	1.4
	1-hydroxybutanone	1.35	1.63
	Total	6.02	7.26
	Sum	43.01	51.83

Conclusion

In this study, we investigated the physical properties of crude bio-oil and bio-oil aqueous phase from switchgrass pyrolysis. Two organic solvents were investigated for chemical extraction from bio-oil aqueous phase. GC/MS analysis for the extracted chemicals indicated ethyl acetate extracted more compounds especially phenolics from bio-oil aqueous phase. Most organic acids and anhydrosugars were remained in the aqueous phase even after organic solvent extraction. Seventeen compounds were quantified by GC-FID and HPLC-PDA using external standards and results indicate that large amount of compounds in bio-oil aqueous phase are acetic acid and levoglucosan.

Acknowledgement

We acknowledge funding for this work from the Department of Energy, BioEnergy Technologies Office under the Carbon, Hydrogen and Separations Efficiency (CHASE) in Bio-Oil Conversion Pathways program, DE-FOA-0000812.

References

[1] Ragauskas AJ, Williams CK, Davison BH, Britovsek G, Cairney J, Eckert CA, et al. The path forward for biofuels and biomaterials. Science. 2006;311:484-9.
[2] Scott DS, Piskorz J. The Continuous Flash Pyrolysis of Biomass. Canadian Journal of Chemical Engineering. 1984;62:404-12.
[3] Mohan D, Pittman CU, Steele PH. Pyrolysis of wood/biomass for bio-oil: A critical review. Energy & Fuels. 2006;20:848-89.
[4] Wei Y, Lei HW, Wang L, Zhu L, Zhang XS, Liu YP, et al. Liquid-Liquid Extraction of Biomass Pyrolysis Bio-oil. Energy & Fuels. 2014;28:1207-12.
[5] Vitasari CR, Meindersma GW, de Haan AB. Water extraction of pyrolysis oil: The first step for the recovery of renewable chemicals. Bioresource Technology. 2011;102:7204-10.
[6] Sipila K, Kuoppala E, Fagernas L, Oasmaa A. Characterization of biomass-based flash pyrolysis oils. Biomass & Bioenergy. 1998;14:103-13.
[7] Oasmaa A, Kuoppala E, Solantausta Y. Fast pyrolysis of forestry residue. 2. Physicochemical composition of product liquid. Energy & Fuels. 2003;17:433-43.
[8] Li HY, Xu QL, Xue HS, Yan YJ. Catalytic reforming of the aqueous phase derived from fast-pyrolysis of biomass. Renewable Energy. 2009;34:2872-7.
[9] Bennett NM, Helle SS, Duff SJB. Extraction and hydrolysis of levoglucosan from pyrolysis oil. Bioresource Technology. 2009;100:6059-63.
[10] Lian JN, Garcia-Perez M, Chen SL. Fermentation of levoglucosan with oleaginous yeasts for lipid production. Bioresource Technology. 2013;133:183-9.
[11] Kechagiopoulos PN, Voutetakis SS, Lemonidou AA, Vasalos IA. Hydrogen Production via Reforming of the Aqueous Phase of Bio-Oil over Ni/Olivine Catalysts in a Spouted Bed Reactor. Industrial & Engineering Chemistry Research. 2009;48:1400-8.
[12] Kechagiopoulos PN, Voutetakis SS, Lemonidou AA, Vasalos IA. Hydrogen production via steam reforming of the aqueous phase of bio-oil in a fixed bed reactor. Energy & Fuels. 2006;20:2155-63.
[13] Kim P, Johnson A, Edmunds CW, Radosevich M, Vogt F, Rials TG, et al. Surface Functionality and Carbon Structures in Lignocellulosic-Derived Biochars Produced by Fast Pyrolysis. Energy & Fuels. 2011;25:4693-703.

[14] Boucher ME, Chaala A, Roy C. Bio-oils obtained by vacuum pyrolysis of softwood bark as a liquid fuel for gas turbines. Part I: Properties of bio-oil and its blends with methanol and a pyrolytic aqueous phase. Biomass & Bioenergy. 2000;19:337-50.
[15] He RH, Ye P, English BC, Satrio JA. Influence of pyrolysis condition on switchgrass bio-oil yield and physicochemical properties. Bioresource Technology. 2009;100:5305-11.

Fast Pyrolysis of Lignocellulosic Biomass by 2-Step Approach for Producing Fractionated Bio-oil Products

Nicole Hammer, John Starcevich, Rene Garrido, Charles Coe, Justinus Satrio

Department of Chemical Engineering Villanova University, Villanova, PA 19085

Lignocellulosic biomass is a promising feedstock for the production of chemicals and fuels. One promising approach that has gained interest for converting lignocellulosic biomass in the past few years is fast pyrolysis. Fast pyrolysis is a thermochemical conversion process in which biomass is decomposed to produce a liquid (bio-oil), a solid (bio-char), and a non-condensable gas mixture (syngas) by exposing biomass to high temperatures, typically between 400 and 600°C, in the absence of oxygen. Fast pyrolysis at the 'standard' 500°C simultaneously converts all three components of lignocellulosic biomass, i.e. cellulose, hemicellulose and lignin. The liquid mixture of bio-crude oil is composed of wide variety of chemical compounds having different physicochemical properties. Since cellulose, hemicellulose, and lignin have different thermal decomposition behavior, the use of a single-step fast pyrolysis at 500°C may not be optimal. Specifically, cellulose and hemicellulose decompose at temperature ranges well below 500°C. Conversions of these biomass components at 500°C result low yield of liquid products since the large molecule organic compounds are further degraded to non-condensable gases and undesirable low molecular weight components.

This research aims at utilizing the decomposition behavior of cellulose, hemicellulose and lignin, to optimize the conversions of each biomass component to liquid bio-oil. This is done by implementing a 2-step fast pyrolysis approach in which biomass is pyrolyzed consecutively twice at two different temperature levels to produce two distinctive bio-oil product fractions. In this process, biomass is first pyrolyzed at a temperature which decomposes mostly cellulose and hemicellulose to produce a bio-oil fraction rich of organic compounds derived from the fibers. In the second step, the remaining solid biomass is pyrolyzed at a temperature which allows optimum conversion of lignin to produce a second bio-oil fraction consisting of predominantly lignin-derived organic compounds. The production of two separate bio-oil fractions, one derived mainly from carbohydrate fraction and the other mainly from lignin fraction of biomass, promises an easier further downstream utilization of bio-oil.

Initial study of 2-step fast pyrolysis of pinewood, cellulose, hemicellulose and lignin using a thermogravimetric analyzer (TGA) revealed significant conversions of hemicellulose and cellulose at the first step up to 70% and 78% at 325°C, respectively. It was observed that when the first step temperature was raised to 350°C, the conversions increased to 73% and 91% for hemicellulose and cellulose, respectively. The conversion of the lignin fraction increased from 18% to 27% when the first step temperature was raised from 325°C to 350°C. Further fast pyrolysis study by using a micropyrolyzer-GC/MS system confirmed that fast pyrolysis of pinewood at 325°C in the first step followed by second step pyrolysis at 500°C produced a sugar-rich bio-oil from the first step and a lignin-rich bio-oil from the second step. This presentation will include recent results from *in situ* catalytic pyrolysis studies using the 2-step fast process.

The influence of biomass components and pretreatments (hot-water extraction and electron beam irradiation) on biomass pyrolysis/ catalytic pyrolysis

Ofei D. Mante[1*], Thomas E. Amidon[2], Arthur Stipanovic[3], and Suresh P. Babu[1]
[1]Department of Sustainable Energy Technologies, Brookhaven National Laboratory, Upton, NY
[2]Department of Paper and Bioprocess Engineering, State University of New York, College of Environmental Science and Forestry, Syracuse, NY
[3]Department of Chemistry, State University of New York, College of Environmental Science and Forestry, Syracuse, NY

Abstract

The poor quality of raw pyrolysis oils and the conversion efficiency of pyrolysis are fundamentally dependent on the characteristics of the starting biomass feedstock. Therefore, a biomass preprocessing step could be used to promote selective pyrolysis of biomass to more desirable components and improve the conversion efficiency. In this investigation, we focused on understanding: 1) the role of biomass components; and 2) the effects of biomass pretreatment (hot-water extraction (HWE) and electron beam (EB) irradiation) on the formation of undesirable oxygenated species during pyrolysis at 550 °C using a pyroprobe-GC/MS system. Unlike many studies where surrogates are used to probe the effect of biomass characteristics on pyrolysis; in this work, we employed well-characterized biomass feedstocks with different natural varying levels of plant cell-wall components. With this approach, possible interaction effects of the biomass components which are not revealed in model studies are captured. Major pyrolysis compounds produced from pyrolysis of debarked sugar maple, poplar, willow, switchgrass, and hot-water extracted sugar maple were correlated with their composition (namely, glucan, xylan, arabinan, mannan, acetyl, lignin, extractives and ash). In addition, we evaluated the effects of HWE at residence times of 0.5 h, 1.0 h, and 2.0 h and EB irradiation at 1000 kGy on pyrolysis of sugar maple. The influence of HWE on the conversion of biomass into aromatic hydrocarbons during catalytic pyrolysis with HZSM5 catalyst was also investigated. Pearson correlation coefficients were used to analyze the data and identify the role of biomass components on production of oxygenated species such as hydroxyacetaldehyde, acetic acid, levoglucosan, hydroxyacetone, methyl pyruvate, furfural, 2-furanone, 5-hydroxymethyl furfural, 4-allyl-2,6-dimethoxyphenol, coniferyl alcohol, and 2-methoxy-4-vinylphenol. The outcome of these studies would be presented.

Natalie Girouard
School of Chemical & Biomolecular Engineering, Georgia Institute of Technology
Dr. Shanhong Xu[1], Dr. J. Carson Meredith[1], and Dr. Meisha L. Shofner[2]

[1] School of Chemical & Biomolecular Engineering
[2] School of Materials Science and Engineering

Abstract

Thermal Stability and Mechanical Reinforcement of Cellulose Nanocrystal/Waterborne Epoxy Composites

One of the building blocks of cellulose structures found in natural materials are individual cellulose crystallites known as cellulose nanocrystals (CNCs). CNCs are reinforcing fillers of emerging interest for polymer composites due to their high modulus and potential for sustainable production. In this study, CNC-based composites with a waterborne epoxy resin matrix were prepared using two different processing strategies and characterized to determine morphology, water content, and thermal and mechanical properties. A major challenge with composite materials is achieving homogeneous nanoparticle dispersion. CNCs have strong intermolecular interactions with each other through hydrogen bonding from hydroxyl groups present on the CNC surface. The surface hydroxyl groups influence the not only the dispersion levels that may be achieved in CNC composites, but also the physical and chemical interactions with the polymer matrix. This study explored the effect of processing by simply altering the addition of mixing on cellulose nanoparticle dispersion and the resulting impact on composite performance. While the level of dispersion achieved was different for the two cases examined in this study, all composites retained the transparency inherent to the neat matrix material. Mechanical and thermal property analyses were correlated with polarized light microscopy in order to explain macroscopic structure-property relationships. Preliminary results have shown that improved nanoparticle dispersion can be achieved through altering the order of addition and without compromising other important variables surrounding composite synthesis. This affords a unique opportunity to compare two systems with varied degrees of nanoparticle dispersion and to assess this impact on bulk composite performance. The results of this study provide evidence that CNCs can improve thermomechanical performance of thermosetting polymers, as observed through changes in the glass transition temperature and modulus values. Overall, the results of this research suggest that CNCs are promising fillers for waterborne epoxy matrices with applications in structural materials and coatings.

370301 Surface Functionalization of Nanocrystalline Cellulose

Abstract:

The chemical modification of nanocrystalline cellulose (NCC) mainly occurs on the surface by introducing functional groups to meet the needs of different applications. The chemical reactions modify the hydroxyl groups of cellulose primarily by esterification, silylation, oxidation and etherification. TEMPO (2,2,6,6-tetramethylpiperidine-1-oxyl) mediated oxidation selectively converts hydroxymethyl group into carboxylic group on the surface of cellulose, which results in a better dispersion of HCl-hydrolyzed NCC. Most of previous efforts on surface functionalization have been focused on the chemical modification of hydroxyl groups of cellulose and improve NCC dispersion in polar organic solvents or its specific mechanical applications. Few researchers have focused on the chemical modification or polymer grafting of NCC to be dispersed in non-polar solvents.

In this study, we will use polyisobutylene succinic anhydride (PIBSA) to graft a hydrophobic tail on the surface of NCC. The succinic anhydride can covalently bond with hydroxyl groups through esterification. In the process, NCC will react with polyisobutylene succinic anhydride (Mw ~950-1300) in toluene at 100°C for 5-30 min. Finally, the resulting nanocellulose-PIBSA will be collected via centrifugation. The grafted NCC will be dispersed in non-polar solvents by mechanical and ultrasonic mixing with the disperser homogenizer. In addition, (2-dodecen-1-yl) succinic anhydride with lower molecular weight will also be used in the polymer grafting to examine the effect of short alkyl chain on the dispersion of NCC.

Injectable hydrogel encapsulated with pH-responsive and amphiphilic copolymer for controlled drug release

*Nusheng Chen[1], *Zhaohui Tong[2],*

[1, 2] Agricultural and Biological Engineering Department, University of Florida, Frazier Rogers hall, Museum Road, PO Box 110570 Gainesville, FL. 32611

*ztong@ufl.edu

Abstract: Injectable hydrogel can work as an effective carrier for drug delivery because of its biocompatibility, minimal invasion and excellent permeability. Furthermore, the entrapment of nanoscopic or microscopic structures inside hydrogel can broaden its application and prolong the delivery of target drugs. In this project, a pH-sensitive copolymer: poly ethylene oxide-block-poly dimethylaminoethyl methacrylate (PEO-b-PDMAEMA) was constructed and embedded into a cellulose derived injectable hydrogel in the form of aqueous micelles. The micelle structure can facilitate the encapsulation of hydrophobic substance and thus, enhance the loading ability for the hydrogel. In addition, the release of inclusion could be triggered by the external stimuli due to pH-sensitive of PDMAEMA. Hydrogel and micelles will then work together as double barriers avoiding the burst release of drugs. In this work, PEO was first modified to become macro initiator and the following ATRP (atom transfer radical polymerization) reaction led to the formation of copolymer PEO-b-PDMAEMA. After dialysis, the corresponding aqueous micelles were generated, which was employed for a hydrophobic dye (drug model) encapsulation. Injectable gel was fabricated following the procedure below. Carboxymethyl cellulose (CMC) was oxidized to induce aldehyde groups and modified to add amine groups respectively. Then, Schiff base reaction rendered the formation of gel by simple mixture of cellulose derivatives under mild condition. The properties of injectable hydrogel were characterized in the aspects of gelation time, elasticity, or swelling ratio. The results showed that the gel could be generated in a very short period after blending. Besides, pH-triggered release characteristic of micelles and its injectable gel complex were studied by fluorescent spectroscopy. The results indicated that pH-sensitive and amphiphilic copolymer not only facilitates the loading of hydrophobic substances, but also provides the capacity of pH triggered release.

389396 Chemical Surface Modification of Cellulose Nanofibrils and Their Reinforcement Effect in Polystyrene Matrix

Abstract:

Cellulose nanofibrils (CNFs) have been of a great interest as reinforcing fillers due to their low weight, high aspect ratio, and excellent mechanical strength properties. The advantages of CNFs, however, are often limited by their aggregation tendency and interfacial compatibility with hydrophobic polymer matrices. Chemical surface modification of CNFs by polymer grafting is introduced to limit the intermolecular hydrogen bonding between nanofibrils by functionalizing surface hydroxyl groups of cellulose and to improve wettability with hydrophobic material while maintain its biodegradability and reinforcing capability. In this study, maleated styrene block copolymers were successfully grafted on the surface of CNFs through esterification. The evidences of grafting were identified through changes in chemical functional groups of CNFs. The influences of polymer grafting on the morphological, surface wetting, thermal stability, crystallinity, and solvent redispersion characteristics were evaluated and compared with those of neat CNFs. The reinforcing capability of modified CNFs as reinforcing fillers in polystyrene matrix was assessed. The study also explored the mechanical performance of different nanocomposite processing methods utilizing neat CNFs. The results of this study would provide indication on the effects of CNFs surface modification as reinforcing fillers in hydrophobic polymers.

RESEARCH ON CRYSTALLINE CELLULOSE HYDROGEN BONDS NETWORK

Ting Cui[1, 2], Jihong Li[1, 2], Menghui Yu[1,2], Shizhong Li[1, 2], Zhipei Yan[1, 2]*
[1] Institute of Nuclear and New Energy Technology, Tsinghua University, Beijing 100084, China
[2] Beijing Engineering Research Centre of Biofuels, Tsinghua University, Beijing 100084, China

Abstract

Background

Hydrogen bonds play an important role in crystalline cellulose's stability. However, crystalline cellulose bioconversion is partly inefficient due to the deleterious impact of cellulose multiple hydrogen bonds on enzymatic accessibility. The type of hydrogen bond (interchain and intrachain hydrogen bond) and the total number of hydrogen bond differ a lot among different cellulose allomorphs. However, there was no systematic report about hydrogen bond network of different crystalline celluloses. Thus, in this study, the hydrogen bond network of crystalline cellulose I, II, III was investigated through Fourier transform infrared spectroscopy (FTIR). The relationship between hydrogen bond network and cellulose enzymatic digestibility was studied.

Results

The results of FTIR indicated that the intrachain/interchain hydrogen bond ratio of crystalline cellulose was strongly correlated with cellulose allomorph. The transformation of crystalline allomorph from native cellulose I to cellulose II, III would led to an increase in the number of interchain hydrogen bonds and a decrease in the number of cellulose intrachain hydrogen bonds. Cellulose enzymatic hydrolysis rate had strongly positive correlation with cellulose interchain/intrachain hydrogen bond ratio. The order of crystalline cellulose interchain/intrachain hydrogen bond ratio was as followed: cellulose III > cellulose II > cellulose I.

Conclusion

In this study, through the experimental method, the results revealed that compared to cellulose interchain hydrogen bond, the intrachain hydrogen bond of crystalline cellulose seemed to be more difficult for cellulase to break. Our experimental efforts may provide a new insight for further improvements of lignocellulose pretreatment method.

Biomass Pyrolysis Reaction Kinetics using Step-Isothermal Thermogravimetry and TGA/MS

Catherine Brewer[1] and Kyriacos Zygourakis[2]

[1] Chemical and Materials Engineering, New Mexico State University, Las Cruces, NM
[2] Chemical and Biomolecular Engineering, Rice University, Houston, TX

ABSTRACT

Several studies have shown that the properties of biochars produced from the same feedstock are controlled by several operating conditions that include the highest heat treatment temperature, heating rate, particle size and heat treatment time at the highest temperature[1]. Since feedstock properties can also vary considerably (even for the same type of biomass), it is clear that the design of reactor systems able to produce biochars with reproducible and tunable properties will require the development of dynamic models that can accurately describe the kinetics of biomass pyrolysis reactions.

This study introduces a systematic method that uses advanced thermal analysis techniques and online mass spectrometry to develop and test a model for the kinetics of biomass pyrolysis reactions. While many earlier studies based their analysis on thermogravimetric experiments with constant heating ramps, our approach begins with step isothermal experiments to identify the potential pseudo-components of our biomass. When studying mixtures of multiple components that decompose at similar temperatures, step isothermal experiments can provide new insights by better separation of the mixture components. The simulated results of Figure 1 demonstrate this principle by heating a two-component mixture at a constant heating rate of 5°C/min (Figure 1A) and according to a step isothermal program (Figure 1B). The two components A and B of this mixture have similar activation energies and will exhibit their maximum decomposition rates at

---

1 Hao Sun, W.C. Hockaday, C.A. Masiello and K. Zygourakis, *Industrial and Engineering Chemistry Research*, **51** (9), 3587–3597 (2012).

closely spaced temperatures. As expected, the constant heating rate produces a differential thermogram (DTG) with a single reaction rate peak, the well-known result of the overlapping rates of our two parallel decomposition reactions. The step isothermal program of Figure 1B, on the other hand, produces two clearly delineated peaks through careful choices of the temperatures and times at which the program switches between the heating ramp and isothermal operation. Moreover, the simulation results show that the volatiles released during the first peak consist primarily of the gaseous products of the first decomposition reaction, while the second peak volatiles come mostly from the second decomposition reaction. Thus, step isothermal methods provide the ideal vehicle for studying the complicated reaction sequences of biomass pyrolysis with online mass spectrometry coupled with thermogravimetry (TGA/MS).

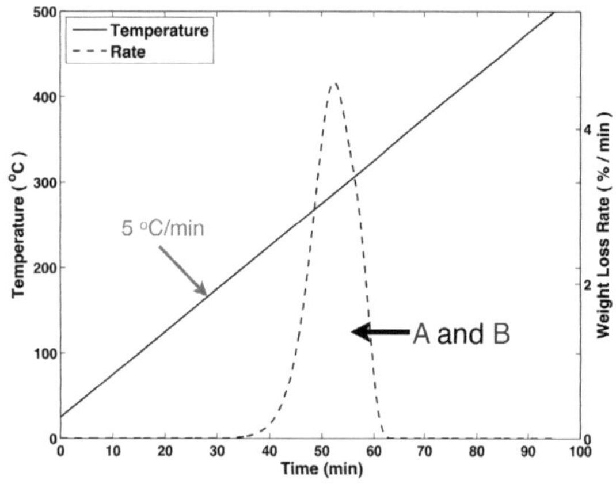

Figure 1: Results from a simulated TGA experiment with a constant heating rate of 5°C/min. Since the activation energies of the two parallel decomposition reactions are close to each other, a single rate peak is observed.

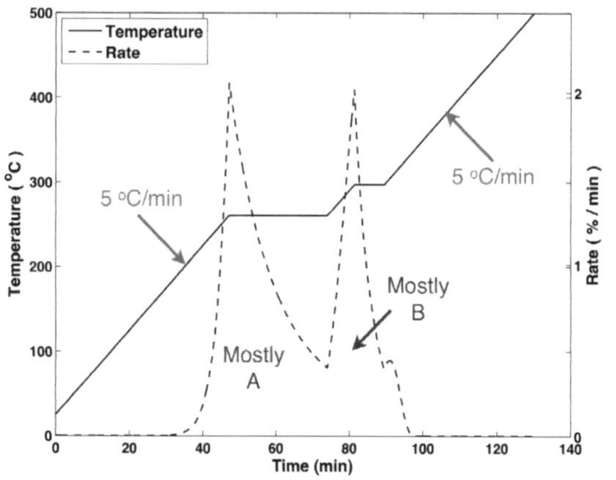

Figure 2: Results from a simulated TGA experiment run with the indicated step isothermal temperature program and the same two-component mixture used in Figure 1. Two peaks are now observed.

Corn stover was chosen to demonstrate this method. We first carried out a series of step isothermal experiments on a TGA/MS system and carefully analyzed the composition of the volatiles emitted at all steps to identify potential pseudo-components that can describe the dynamic behavior of our pyrolyzing biomass. Dynamic simulations and nonlinear least-squares methods were then used to test several reaction networks and select the network that can best fit the experimental data. The selected network was further tested using dynamic TGA/MS experiments with a variety of temperature programs (using both constant heating ramps and high-resolution TGA techniques).

Our results showed that a network of first-order parallel and sequential reactions with as many as five solid pseudo-components must be used to accurately describe the pyrolysis dynamics of corn stover. The kinetic parameters of the individual first-order reactions and their heats of reaction were also estimated using thermogravimetric and differential scanning calorimetry data.

Finally, we validated the kinetic model using a well-characterized library of fourteen biochars that have been produced using a fixed bed reactor, seven different final pyrolysis temperatures and two different atmospheres: (a) pure nitrogen and (b) a mixture of 5% oxygen in 95% nitrogen. TGA/MS results from step isothermal experiments with these chars clearly showed that a dynamic model consisting of parallel and sequential reactions must be employed to match the properties of the produced biochar to their temperature history.

Catalytic Enhancement of Alkaline Thermal Treatment of Cellulose to H_2 in the presence of $Ca(OH)_2$ and $Mg(OH)_2$

Maxim Stonor[2], Thomas Ferguson[1], Jingguang Chen[2], and Ah-Hyung Alissa Park[1,2]

[1]Department of Earth and Environmental Engineering
[2]Department of Chemical Engineering,
Columbia University in the City of New York, New York 10027

As concerns over climate change continue to grow, a shift away from fossil fuels and to a sustainable energy source such as biomass is highly desired. Particularly, bioenergy with integrated carbon capture and storage would allow not only carbon neutral but also carbon negative energy conversions. In this study, the production of H_2 from biomass such as cellulose is achieved by reacting it with strong base under moderate reaction temperature and pressure. This novel biomass conversion pathway fixes carbon into the solid matrix as carbonates, and thus, the product gas is free of COx. The regeneration of hydroxide and subsequent storage of CO_2 would be important for the overall sustainability of the technology. NaOH has been investigated as the main source of the strong base, and other hydroxides, such as $Ca(OH)_2$ and $Mg(OH)_2$, are being considered as more sustainable alternatives for NaOH. The reactions of these weaker hydroxides with cellulose resulted in lower conversion to H_2 (1% and 0.3%, respectively) when compared to NaOH (33%). In order to promote the activity of these systems with weaker hydroxides, a 10% Nickel on Zirconia catalyst was employed at 20% sample loading giving the $Ca(OH)_2$ reaction a yield of 29% and $Mg(OH)_2$ 16%. The distribution of other product gases (i.e., CH_4, CO and CO_2) also varied with the type of hydroxide used: $Mg(OH)_2$, $Ca(OH)_2$ and NaOH.

2014 AIChE Annual Meeting
Nov. 16 -21, 2014
Atlanta, GA
Session: T4A048 Thermochemical Conversion of Biomass I

Warm Syngas Cleanup Using Carbon-Based Sorbents

Jiajia Meng, Patrick Woolcock, Tim Hansen, Kevin McCabe, Andrew Lucero, Santosh Gangwal*

Southern Research Institute, Environment and Energy Center, Durham, NC 27712

Email: gangwal@southernresearch.org; Phone: 919-282-1053

Tar and particles in syngas produced from gasification of biomass or waste materials need to be removed before fuel production or power generation using an engine. There are, however, no satisfactory tar removal processes to date in terms of simplicity of use and economics. This is because tar in syngas consists of numerous organic constituents including refractory organic compounds; such as polyaromatic hydrocarbons (PAHs) that do not crack in short residence time at temperatures as high as 850 °C even in the presence of expensive catalysts, and form difficult to capture aerosols at low temperatures. As an alternative, high surface area carbon based sorbents can potentially adsorb the tar and capture the particles at warm gas temperatures. We have investigated a process concept based on carbon based sorbents for warm syngas cleanup using a lab-scale fixed bed adsorption reactor. A model tar compound, toluene, mixed with nitrogen and syngas were employed to test the adsorption performance of various carbon-based sorbents. The effect of adsorption conditions on toluene adsorption breakthrough time was investigated using factorial experimental design and optimized conditions were determined. Results indicated a strong correlation between surface area of carbon and adsorption capacity. The results from the study were used to design a pilot-scale adsorption reactor for clean syngas generation from a 3 ton per day pilot-scale biomass gasifier.

390328 Parametric Gasification Study Comparing Traditional and Blended Feedstocks at Varying Conditions

Abstract:

The sustainability and costliness of traditional biomass feedstocks are of concern for thermochemical biomass conversion processes. To be competitive with oil-derived fuels, it is necessary for biomass feedstocks to be available year-round without interruption. Currently, no conventional feedstock can supply a high-throughput, continuous thermochemical biomass conversion process year-round. Our work focuses on comparing the gasification product slate of blended feedstocks, prepared by Idaho National Laboratory, with conventional woody and herbaceous feedstocks. We gasified all feedstocks in a 10.2 cm i.d. gasifier with a continuous solid feeder, hot char removal, and a dodecane scrubber for vapor condensation and separation. Gas chromatography (GC), non-dispersive infrared spectroscopy (NDIR), and a diode laser were used to speciate and quantify non-condensable gases, hydrogen sulfide, and ammonia in the raw syngas stream. Molecular beam mass spectrometry (MBMS) was used to measure and quantify tar species in the hot vapor product stream (~425°C). We compared the gasification products of all feedstocks at 750°C, 800°C, and 850°C. Our goal for this work is to determine if blended feedstocks can be tuned to take on characteristics of traditional feedstocks. If this is achievable, then less desirable feedstocks such as wood residues and whole wood can be blended with other feedstocks to create an optimal blendstock.

379432 Experimental Determination of the Structural Evolution of Biomass Particles during Gasification

Abstract:

Gasification has been identified as an energy-efficient, environmentally-friendly and economically-feasible technology to partially oxidize biomass into a gaseous mixture of syngas consisting of H_2, CO, CH_4 and CO_2. High-quality syngas can be further used to catalytically synthesize liquid fuels and produce hydrogen. Biomass gasification involves complicated structural evolution of solid biomass particles. During gasification, the physicochemical structure of biomass particle changes significantly when the volatiles are released. The change of the porous structure of biomass particles during gasification is critical in determining the mass transport of volatile species, and the physical properties of the particle. This research is to measure the structural evolution of biomass particles during gasification. A specifically designed tubular reactor and a thermogravimeter (TGA) are used to prepare biomass samples pyrolyzed or gasified at different temperatures. The porous structure, morphology, and structural evolution of biomass particle are analyzed by a surface area/pore analyzer, scanning electron microscopy (SEM), transmission electron microscopy (TEM), X-ray diffraction analysis (XRD), Infrared (IR) and Raman spectroscopy. The porous structure of the biomass particles will be first studied by using N_2 adsorption at 75 K using a Micromeritics ASAP 2020 surface area and pore analyzer. The pore structure of the biomass particles including pore size and pore size distribution as well as surface area are determined by the Brunauer-Emmett-Teller (BET) equation. Direct observation of the porous structure is performed using a Carl Zeiss Auriga-BU FIB FESEM microscope and Carl Zeiss Libra 120 Plus TEM microscope. The results from XRD and Raman examination will reveal carbon structure evolution during the gasification process.

378462 Hydrodeoxygenation of 2-Furyl Methyl Ketone As a Model Compound in Bio-Oil from S. japonica Pyrolysis in Fixed-Bed Reactor

Abstract:

The model ketone of bio-oil, *2-furyl methyl ketone* (FMK) or 1-(2-furanyl)-ethanone, which is the main component in *Saccharanina (Laminaria) Japonica (S.Japonica)* pyrolysis products, was successfully upgraded into methyl cyclopentane in the liquid phase and methane in the gas phase. Hydrodeoxygenation (HDO) of FMK over cobalt based phosphide supported on γ-Al_2O_3 substrates made the conversion possible and highly effective. Complete HDO of FMK into methyl cyclopentane and methane was achievable at 400 °C reaction temperature using CoP/γ-Al_2O_3 catalyst synthesized with 10 wt.% Co loading and calcined at 500 °C.

378577 Fast Pyrolysis of S.japonica in a Bubbling Fluidized Bed Reactor for Bio-Oil Production

Abstract:

Recently, macro-algae are recognized as one of the potential feedstock for renewable energy and fuel production. *Saccharina Japonica* is a kind of macro-algae, which is extensively cultivated in Korea for the production of value added chemicals and fuel. *S. Japonica* can be converted into bio-oil, gas and char through fast pyrolysis. In this study, fast pyrolysis of *S.Japonica* was performed in bubbling fluidized-bed reactor under various reaction conditions (pyrolysis temperature, flow rate of fluidizing medium, biomass particle size) to investigate the effect of these variables on the product yields and the qualities of bio-oil. The results showed that the highest yield of bio-oil was 40.87%, obtained at pyrolysis temperature of 350 °C with gas flow rate of 9.6 L/min, feeding rate of 100 g/hr. The biomass particle size also affected the bio-oil yield. The bio-oil contained highest selectivity for propanone and di-anhydromannitol were 43.93% and 22.19%, respectively. The ^{13}C NMR showed the shorter aliphatic carbons and lower molecular weight compounds of alcohol, carbohydrate and ketone were generated during pyrolysis at higher temperature, due to secondary decomposition reactions. The collected biogas after reaction was analyzed by Gas Chromatography with FID detectors. The compositions of bio-char were determined by elemental analysis and inductively couple plasma.

378583 Fast Pyrolysis of Hemp in a Bubbling Fluidized Bed Reactor

Abstract:

Biomass has been focused as one of the alternative energy sources since it is renewable, abundant worldwide, eco-friendly, and carbon neutral. Hemp is a kind of biomass which can be easily cultivated with rapid growth rate. Also, a lot of hemp by-products are produced from textile industry. In this study, hemp has been studied to understand pyrolysis reaction characteristics and fast pyrolysis to evaluate the efficiency of bio-energy production. Pyrolysis kinetics of hemp by-products was studied using a micro tubing reactor. Simple pyrolysis mechanism was proposed and the reaction rate constants were calculated. Hemp by-products were fast pyrolyzed in a bubbling fluidized bed reactor at various reaction conditions. The effects of pyrolysis temperature between 440 °C and 560 °C on product yields were investigated. The yield of bio-oil was changed between 16.11 wt% and 39.44 wt%, and showed higher bio-oil yield with increasing reaction temperature. Aromatic compounds such as furfural, 1.2-benzenedial, 2.3-dihydro-benzofuran and phenol derivatives in bio-oil were major products by the analysis of GC-MS.

385817 Integrated Low Temperature Ionic Liquid Pretreatment of Biomass

Abstract:

The high crystallinity of cellulose of native lignocellulosic biomass, while imparting structural integrity and mechanical strength to the material, renders it recalcitrance towards hydrolysis aimed at producing glucose. Neither water molecules nor catalysts for hydrolysis (saccharification) are able to easily penetrate the crystalline matrix. Room temperature pre-processing is coupled with ionic liquid incubation (IL) at low temperatures ranging between 30 and 50°C, producing an easily hydrolysable cellulosic substrate at dramatically lower incubation temperatures. Moreover, by altering the IL displacement solvents, this method can be tuned to retain or reject biomass components from the IL phase, providing a method for separation of hexose and pentose sugars.

397478 Chemistry and Kinetics of Biomass Gasification Using Carbon Dioxide

Abstract:

In recent years, gasification has originated as one of the better techniques to harness the energy from biomass. Syngas produced during gasification can be used as a feedstock in industrial applications for electricity, chemicals and fuel production. Air, oxygen and steam have been widely used as oxidizing medium in the gasification process. The use of Carbon dioxide (CO_2) as the oxidizing agent in biomass gasification has been explored in limited studies. Catalytic gasification of biomass can effectively improve the reaction rate and conversion efficiency compared to conventional gasification. However, very little is known about the catalytic gasification of biomass char with CO_2.

In this study, a widely accepted model for char gasification, the random pore model (RPM) is to be fitted onto data obtained from the experiments. Initially, unloaded biomass will be subjected to gasification using CO_2. Using the evolution of CO in the gasification as basis for conversion, the values of parameters of the model are calculated. The biomass is then loaded with potassium and calcium catalysts and is to be subjected to the same procedure. The effect of the catalysts on these parameters is to be observed.

In a previous study, the chemistry of the biomass gasification using CO_2 was studied by operating the process at different temperatures in a bench scale fluidized bed reactor. The output syngas was analyzed and based on the trends of the component gases, a set of reactions dominating the process was postulated. A number of gas analysis instruments such as a GC/MS (Agilent 7890/5975), a micro GC (Agilent 490), a portable gas analyzer (Nova 7905AQN4), and an FTIR are used to collect necessary information.

396059 Separation of Cellulosic Biomass from Ionic Liquids using Compressed CO2

Abstract:

The rigid hydrogen bonding network between cellulose molecules renders it insoluble in most organic solvents. However, select ionic liquids (ILs) including 1-ethyl-3-methylimidazolium diethyl phosphate [EMIm][DEP] are capable of dissolving significant quantities. Through IL pretreatment cellulose crystallinity is reduced thus enhancing reactivity. As pre-processed cellulose is non-volatile, anti-solvent techniques are most often used for the separation from ionic liquids. We have recently discovered that compressed carbon dioxide presents a unique alternative to conventional organic/aqueous antisolvent methods. Phase equilibrium, high pressure NMR, and solid state NMR results will demonstrate characteristics of the CO_2 based physical (non-reactive) separation process and the amorphous cellulose product.

394788 Alkaline Pretreatment of Corn Stover: Pilot-Scale Fractionation, Stream Characterization, and Bench-Scale Enzymatic Hydrolysis

Abstract:

Alkaline pretreatment is a process by which lignocellulosic biomass is separated into a carbohydrate rich solid phase and a lignin rich liquor phase using a caustic aqueous solution. Previous bench scale work has shown that the efficacy of the alkaline pretreatment depends upon contact time, temperature of the solution, severity, and constitution of the biomass. A three factor, two level central composite experimental design with triplicate runs at the center point was conducted at the pilot scale to determine the relationship of pretreatment NaOH loading, temperature, and AQ loading on solids solubilization and component yields. Alkaline pretreatment was performed in a 130 L, jacketed, horizontal vessel with agitation at 15 rpm for 30 minutes at 10% (w/w) TS. In addition to the 17 alkaline pretreatment runs conducted within the central composite experimental design, a wash step was applied to five additional pretreatment runs to examine the effect of washing on lignin removal, alkaline pretreated insoluble solids composition, and enzymatic hydrolysis yields. This presentation will address the procedures for scaling this process to pilot scale as well as resulting pretreatment yields and enzymatic hydrolysis results.

UPGRADING LIGNIN ANTIOXIDANT PROPERTIES BY SELECTIVE DISSOCIATION OF WHEAT STRAW LIGNIN LINKAGES

Mohammadali Azadfar and Shulin Chen

Department of Biological Systems Engineering, Washington State University, Pullman, WA 99164-6120, USA

Funding Sources: Funding for this research was provided by Washington State University

ABSTRACT

Production of value-added products from lignin obtained after pretreatment of biomass requires advanced characterization and modification techniques to illustrate the lignin structure in detail and also to increase the lignin's usability. The lignin modification methods which have been proposed so far endeavor to decompose the natural structure of lignin for producing the value-added products. One of the attractive applications of raw lignin obtained after pretreatment is for use as an antioxidant. In order to preserve the intrinsic structure of lignin for producing antioxidant, the lignin linkages should be dissociated utilizing the ingenious methods. This project aims to modify the lignin structure employing the benefits of the effect of alkaline and acidic conditions on ester and ether lignin linkages without destruction of valuable original structure of wheat straw lignin.

The characterization of wheat straw lignin obtained from sequential ozone and soaking aqueous ammonium pretreatment revealed the presence of phenolic derived-compound with comparable antioxidant activity with other synthetic antioxidants. The characterization of lignin structure demonstrated that the good antioxidant activity of the lignin is due to the predominant presence of hydroxycinnamic acids and G-lignin derived-compounds such as ferulic acid and guaiacol during the alkaline pretreatment (S/G ratio of 0.35). Alkaline pretreatment (soaking aqueous ammonium in this project) of wheat straw is mainly responsible for breaking down of the ester linkages. The dissociation of ester linkages open this opportunity that the hydroxycinnamic acids derivatives can be appeared with applied antioxidant properties including the presence of hydroxyl group, *ortho*-methoxy substitution on the phenolic ring, and low molecular mass (<1000 Mn).

Regarding the molecular mass of lignin's precursors and characterization of obtained lignin structure, it is possible to determine the number of linked lignin monomers (G, S, H lignin units). Consideration of the lignin macromolecule structure opens this opportunity to characterize the potential linkages among lignin monomers. The ether linkages have predominant capability to link the lignin's monomers in order to show the obtained lignin's molecular weight. Based on the selective destruction effect of acid on ether bonds, it has been proposed that by utilizing the acidic condition the majority of ether linkages in lignin structure will be dissociated. Thus, lignin-derived compounds with a substantial lower molecular weight and also containing new active functional groups will be obtained. The prepared lignin derivatives including aforementioned properties will be indicated excellent antioxidant activity.

Keywords: Lignin linkages; Selective dissociation; Characterization, Modification; Antioxidant

Engineered Kraft Lignin Particles for Foam Stabilization

Stephanie Lam, Andrew Tibbits, Orlin D. Velev

Department of Chemical & Biomolecular Engineering, North Carolina State University, Raleigh, North Carolina 27695-7905

Lignin, which exists in the cell wall of woody plants, is one of the most abundant biopolymers found in nature and has many potential uses as a stabilizer, encapsulation agent, and as a component of composite materials. In this work, we present the results of a process where KL particulates, which are themselves emulsifiers, are reformed into drastically more efficient foam stabilizers using a water-based shear process. The flash-precipitation procedure is environmentally-friendly and does not involve the use of organic solvents or harsh chemicals. We characterized the effect of this reformation process on the surface chemistry and structure of the nanoscale voluminous lignin particles produced. The ζ-potential of the KL particles was found to change with pH, and became slightly lower following the reforming process. The effective volume of the particles was found to increase by ~30 times following the reformation procedure, which resulted in an enhanced ability of reformed KL to sterically stabilize foams and emulsions. We also evaluated the stability of foams made with reformed particles at various pH conditions and found enhanced foam stability to correlate with KL reformation at low pH (~2), which corresponds to a surface potential of approximately -20 mV and increased particle hydrophobicity. Thus, we have developed an efficient, environmentally-friendly method by which to expand the effective volume of lignin particles, which after the reformation process have shown enhanced properties as foam and potentially emulsion stabilizers.

2014 AIChE Annual Meeting
Division 17: Forest and Plant Bioproducts Division
17004 Biobased Materials III: Lignin-Based Materials
November 16-21, 2014
Atlanta, GA

Charaterization of Biochar from Loblolly Pine: Changes in Carbon Structure and Consequent Effects on Applications as Sorbents for Aromatic Contaminants

Junyeong Park [1,*], Kwang Hun Lim [2], Martin A. Hubbe [1], Orlando J. Rojas [1], Sunkyu Park [1]

[1] Department of Forest Biomaterials, North Carolina State University, Raleigh, NC 27695
[2] Department of Chemistry, East Carolina University, Greenville, NC 27858
* Presenting author: jpark7@ncsu.edu

Thermally-treated biomass has been generally considered as lower-valued solid fuel or by-products from various thermal conversion processes for lignocellulosics. Recently, however, these solid products such as biochars and their activated counterparts, are getting wide attentions, due to their possibilities for sustainable and versatile applications, including soil amendment, carbon sequestration, and sorbents for various environmental contaminants. It has been reported in existing studies that the structural characteristics of thermally-treated biomass have correlations with required properties for those applications. However, the details about what is going on inside lignocellulosic biomass during thermal treatment have not been fully understood yet. In this study, the changes in structural characteristics, such as aromatic structures and sizes of carbon clusters, were monitored using several complementary analytical tools, including solid-state NMR spectroscopy, compositional analysis, and other characterization methods. After the thermal treatment of loblolly pine at 700 °C under inert atmosphere, the fraction for aromatic carbons inside was increased to 69 % out of total carbons in solids, and even higher to 76% after the subsequent activation process. When biomass was treated under slightly oxidative atmosphere, the development of aromatic cluster was promoted substantially. Surface analysis revealed that the properties like BET surface area and pore volume were mainly dependent on the structural characteristics of the feedstock before activation. It was shown that structural and surface characteristics of thermally-treated biomass were largely governed by process conditions for thermal treatment, and this can make it possible to design solid products for specific applications with enhanced properties.

390387 Phase Behavior of Kraft Liqnins with a Tunable, Renewable Solvent System

Abstract:

Lignin is unique among biopolymers in having significant aromatic character, which makes it potentially attractive for a wide range of uses, including polymeric applications and carbon fibers. Unfortunately, most of the commercial-grade lignins available today (i.e., the Kraft lignins) are severely limited in their applications because of their high metals (primarily sodium) content.

However, recently Clemson researchers have discovered a powerful, versatile, and renewable solvent system that creates a phase behavior when contacted with lignin such that the sodium content can be reduced 10-fold with each phase split; furthermore, this phase separation also separates the lignin into fractions with significantly different molecular weights and chemical properties. Thermodynamic variables such as temperature, solvent composition and amount, and lignin type (i.e., hardwood vs. softwood) have been studied to determine their effect on the location of phase boundaries, the metals selectivities between phases, and the molecular weight distribution of the lignin within each phase.

Pseudo-ternary phase diagrams will be presented for different types of feed lignins with the solvent system at several different temperatures, and the global phase-behavior characteristics of this lignin-solvent system will be discussed, along with the implications of the observed phase equilibria for the development of a viable purification and fractionation process for lignin.

392452 Production and Evaluation of Bioplastic Blends from Phbv, PCL, Epoxidized Natural Rubber and Lignin

Abstract:

Biodegradable polymers, from biological sources such as as poly(3-hydroxybutyrate-co-3-hydroxyvalerate) (PHBV) and petroleum based sources such as poly(ε-caprolactone) (PCL), have been shown to be effective in solving many of the problems with current commercial plastics, including greenhouse gas emissions, continued stability of fossil fuels and concerns related to end of life. Lignin, a by-product of the paper and cellulosic ethanol industries, has recently received attention in biomaterials and chemical applications due to its favorable polyphenolic structure and large quantity of available functional groups. Because of the widespread nature of the pulp and paper industry, the most commonly available form of lignin is the kraft lignin, a byproduct refined from the black liquor stage of the pulping process. The main challenge associated with composites of biopolymers and lignin the appearance brittle behavior caused by the presence of lignin in the polymer matrix. In this study small quantities of epoxidized natural rubber (ENR) have been compounded with the bioplastic blends to operate as an impact modifying agent. Characterization of these bioplastic blends was performed through mechanical testing (tensile, flexural test and impact tests) and thermal properties of the bioplastic blends. The emphasis of this study was improving the impact resistance of lignin-biopolymer blends.

The Researchers acknowledge the financial support from the Natural Sciences and Engineering Research Council (NSERC), Biomaterials and Chemicals Network.

389655 Physicochemical Properties of Upgraded Bio-Oils from Catalytic Pyrolysis of Biomass: A 13C NMR Spectroscopic Investigation to Understand the Effects of the Chemical Composition of the Oils

Abstract:

The production of infrastructure-ready biocrude oil from catalytic pyrolysis offers an opportunity to co-process biomass derived intermediates alongside petroleum feedstocks such as atmospheric gas oil (AGO) and vacuum gas oil (VGO) in standard refinery units. Recent studies [1-4] suggest that, up to 20wt% of upgraded bio-oils can be directly cracked in fluid catalytic cracking (FCC) process with these petroleum feedstocks without affecting the yields. Nevertheless, one of the critical success factors for commercial scale bio-oil co-processing is the physicochemical properties of the upgraded bio-oil. Since the upgraded bio-oil produced in catalytic pyrolysis is not completely free of oxygenated species, there is the need to fundamentally understand the impact of its chemical composition on the physicochemical properties. In this work, several upgraded bio-oils with varying fuel qualities produced from catalytic pyrolysis of different biomass feedstocks with HZSM-5 at 475 ^{o}C were characterized. The kinematic viscosities ranged from 4 to 60 cSt; the pH values were between 3.2 and 5.0; the storage stability varied from 0.02 to 0.84 cSt/day; the densities were within 0.96 and 1.14 kg/m^3, the O/C molar ratios ranged from 0.1 to 0.28; and the nitrogen contents were between 0.1 and 2.12 wt%. Using ^{13}C NMR spectroscopy and Pearson correlation coefficients, we were able to establish relationships between the various functional groups and the following physicochemical fuel properties: long-term storage stability, acidity, viscosity, density, solid residue, and elemental composition. The evidence from this investigation suggests that residual oxygenated species such as phenolics, anhydrosugars, and carbonyls continue to negatively impact the fuel properties of even upgraded bio-oils with low oxygen content levels (e.g.,12 wt.%.). In this talk, we will also discuss the opportunities and challenges of utilizing biocrude oil as a feed for standard refinery units.

References

1. Agblevor, F.A., O. Mante, R. McClung, and S.T. Oyama, *Co-processing of standard gas oil and biocrude oil to hydrocarbon fuels.* Biomass and Bioenergy, 2012. **45**(0): p. 130-137.

2. de Miguel Mercader, F., M.J. Groeneveld, S.R.A. Kersten, N.W.J. Way, C.J. Schaverien, and J.A. Hogendoorn, *Production of advanced biofuels: Co-processing of upgraded pyrolysis oil in standard refinery units.* Applied Catalysis B: Environmental, 2010. **96**(1-2): p. 57-66.

3. Fogassy, G., N. Thegarid, Y. Schuurman, and C. Mirodatos, *The fate of bio-carbon in FCC co-processing products.* Green Chemistry, 2012. **14**(5): p. 1367-1371.

4. Thegarid, N., G. Fogassy, Y. Schuurman, C. Mirodatos, S. Stefanidis, E.F. Iliopoulou, K. Kalogiannis, and A.A. Lappas, *Second-generation biofuels by co-processing catalytic pyrolysis oil in FCC units.* Applied Catalysis B: Environmental, 2014. **145**(0): p. 161-166.

384346 The Importance of Reducing End Functionality in Oligosaccharide, Anhydrooligosaccharide and Cellulose Pyrolysis

Abstract:

Introduction:

Fast pyrolysis of biomass offers the potential for generation of fuels and valuable chemicals from renewable resources in a relatively simple and scalable manner. In order to gain control over the product distribution obtained from fast pyrolysis of biomass, a deeper understanding of the sequences and mechanisms of the reactions occurring during fast pyrolysis is needed. Current fast pyrolysis methods generate product distributions which are the result of several steps of both primary and secondary reactions. In order to better understand the origins of the final pyrolysis product distribution we utilized a set of pyrolysis / tandem mass spectrometry methods (developed in house) in which the initial products can be detected and characterized.

Methods:

The pyrolysis products were formed via direct solid-solid contact on a resistively heated platinum ribbon that was heated at a rate of 1000 $K \times s^{-1}$ to 873 K. After they evaporated from the ribbon, they were quenched to near room temperature and ionized. Two ionization methods were used to ionize the pyrolysis products, positive mode Atmospheric Pressure Chemical Ionization (APCI) with ammonium attachment, and negative mode APCI with chloride attachment. These methods are complementary in that although many molecules can be ionized using both, there are a few exceptions, namely furan-related molecules, which were only ionized via protonation in the positive mode ammonium attachment method.

Results and Discussion:

Using a combination of a fast-heating pyroprobe and a linear quadrupole ion trap mass spectrometer, we have demonstrated that fast pyrolysis of cellulose yields only a few primary products, about half of which have not been reported in literature. These previously unreported pyrolysis products of cellulose contained several products with a ring-opened fragment attached to one or more intact glucopyranosyl rings. Additionally, pyrolysis of oligosaccharides of different chain lengths was performed (2 through 6 units). Remarkably, the major products observed upon fast pyrolysis of cellobiose were also observed for cellotriose, cellotetraose, cellopentaose and cellohexaose albeit with different relative abundances. In order to better understand the origin of the ring opened products observed from oligosaccharide pyrolysis, a range of different carbon labeled cellobioses were examined ([1-^{13}C]glucopyranosyl-glucose, glucopyranosyl[1-^{13}C]glucose, glucopyranosyl[3-^{13}C]glucose, glucopyranosyl[5-^{13}C]glucose). New mechanisms were proposed to explain the ring opened products observed from cellobiose pyrolysis based on these labeling results and were evaluated computationally using Density Functional Theory (DFT) performed at the M06-2X/6-311++G(d,p) level of theory. The newly proposed mechanisms represent a feasible low energy pathway to explain the formation of the ring opened products.

It was also observed that the relative abundance of products with an intact ring attached to a ring-opened fragment decreased with increasing chain length, which suggests that these products are associated with the ratio of terminal to total glucose units. The product distributions produced from pyrolysis of these oligosaccharides differs from cellulose in that they produce a higher relative abundance of products that are the result of ring fragmentation than cellulose. Furthermore, the largest product produced from cellotriose through cellohexaose with a relative abundance greater than 10% was cellotriosan. Inspired by this observation, fast pyrolysis of cellotriosan was also performed, which is noteworthy for its lack of a reducing end compared with standard glucooligosaccharides, which prevents tautomerization of the terminal unit into the ring opened form. Fast pyrolysis of cellotriosan resulted in a product distribution that was nearly identical to that from cellulose pyrolysis. Delineation of the fast pyrolysis reaction pathways and mechanisms for cellotriosan may prove extremely valuable for gaining control over the product distribution from fast pyrolysis of cellulose. Our results indicate that cellotriosan presents an excellent small-molecule surrogate for further experimental and theoretical studies on fast pyrolysis of cellulose.

Low-Order Modeling of Biomass Particle Mixing and Reaction in a Bubbling-Bed Fast Pyrolysis Reactor

Gavin Wiggins (ORNL)
Stuart Daw (ORNL)
Jack Halow (Separation Design)

AIChE 2014 Annual Meeting
Thursday, Nov. 20, 2014

Computational Pyrolysis Consortium
http://energy.ornl.gov/cpc

Separation Design Group
Technology for a Sustainable Energy Future

OAK RIDGE National Laboratory

Background

The work presented here is part of the Computational Pyrolysis Consortium (CPC), a multi-lab activity sponsored by DOE's Bioenergy Technologies Office (BETO).

The main goal of the CPC is to develop computational tools that support the assessment of advanced catalytic technologies for producing infrastructure compatible transportation fuels from biomass-derived pyrolysis oils.

OAK RIDGE National Laboratory | NREL National Renewable Energy Laboratory | Pacific Northwest NATIONAL LABORATORY | INL Idaho National Laboratory | Argonne NATIONAL LABORATORY

Computational Pyrolysis Consortium

2

Fast pyrolysis of biomass

Pyrolysis involves rapid heating of biomass in the absence of air or oxygen to produce noncondensable gases, solid char, and liquid.

Goal of **fast pyrolysis** is to maximize the production of liquid yield (a.k.a. bio-oil or tar).

The liquid can be stored and transported, and used for energy, chemicals or as an energy carrier. [Bridgwater 2012]

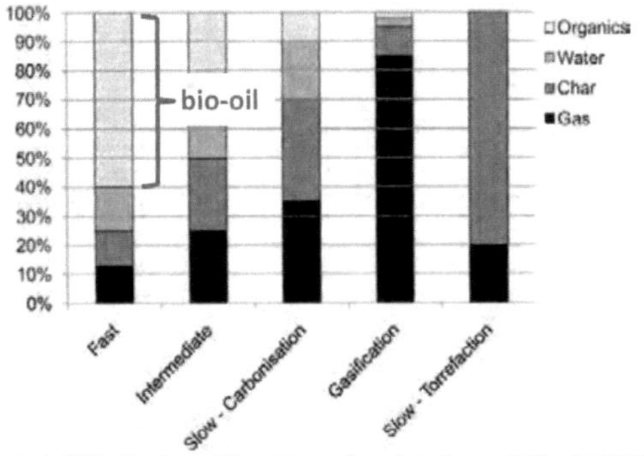

Product distribution from different types of pyrolysis, *Source: Bridgwater 2012*

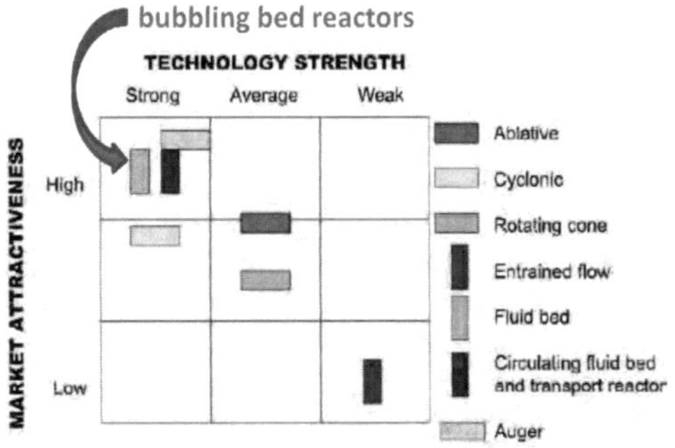

Commercial potential of fast pyrolysis technologies, *Source: Butler 2011*

3

Bubbling bed reactors are widely used for fast pyrolysis

- Geldart Group B (sand) bed particles with or without catalysts
- Bed fluidized under no-oxygen conditions
- Raw biomass injected as particles and removed as char
- Biomass typically < 1% of bed mass
- Bed temperature 425-600°C
- Very **rapid heating rate** (up to 1,000°C/s)
- **Mixing and particle RTD** very important to product composition and conversion

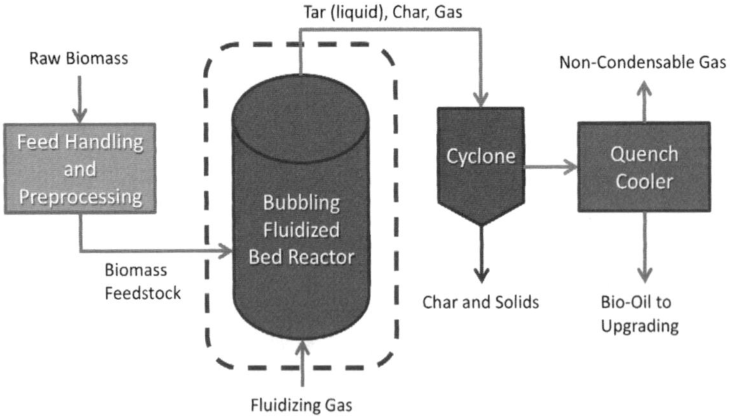

4

Objectives

- Create low-order reactor models (LORM) with the following characteristics
  - fast-executing (can be run on a desktop computer)
  - open-source tool for the broader pyrolysis community
  - account for rate-limiting steps
    - particle-scale mixing
    - heat and mass transfer
    - secondary reactions (including catalytic) of the released pyrolysis gases

- Configure those models to facilitate rapid screening of process and scale-up options for industry and research community

5

Modeling Approach

Create the following models:

■ focus of today's discussion

- Sub-model for intra-particle biomass heat-up and reaction
- Statistical biomass particle mixing sub-model
- Gas-phase reaction/convection sub-model for released gases
- Diffusion/boundary layer sub-model for pyrolysis products
- Global reactor model to integrate the sub-models

Integrated Low-Order Reactor Model (LORM)

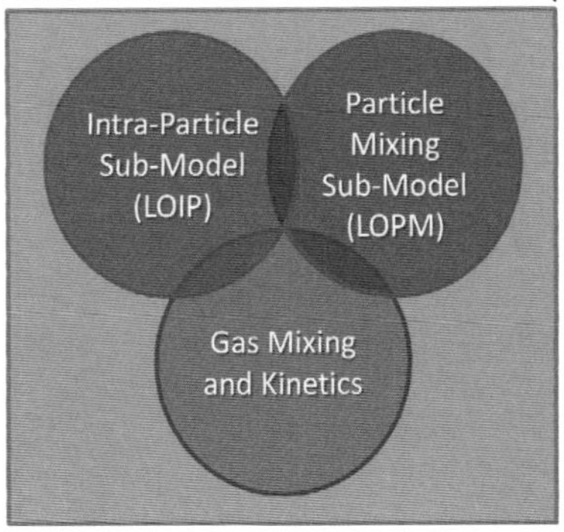

6

Biomass Particle-Scale Modeling Challenges

Literature on intra-particle kinetics is inconsistent , typically available for limited experiments.

[Chan 1985, Di Blasi 1993, Babu 2003, Gronli 2000, Kersten 2005, Prakash 2008, Shafizadeh 1982]

Available pyrolysis models treat biomass particles as "one" size, but grinders and mills often produce broad distributions of sizes.

[Di Blasi 2002, Bryden 2002, Chaurasia 2003, Cui 2007, Galgano 2003, Galgano 2004, Gronli 2000, Haseli 2011, Janse 2000, Koufopanos 1991, Kung 1972, Larfeldt 2000, Miao 2011, Papadikis 2009]

Particle surface boundary conditions are highly complex. [Papadikis2010]

Biomass is typically anisotropic and inhomogeneous with wide variations among different species.

[Chaurasia 2003, Babu 2004, Gronli 2000, Haseli 2011, Koufopanos 1991, Kung 1972, Larfeldt 2000, Okekunle 2011, Papadikis 2010, Prakash 2009, Pyle 1984, Sadhukhan 2009]

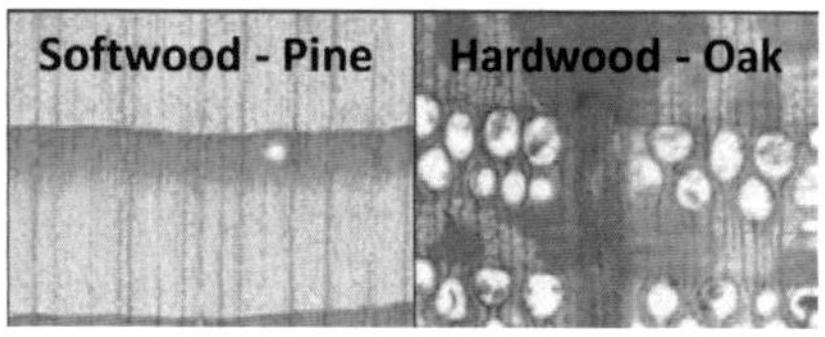

Source: Wood Handbook 2010

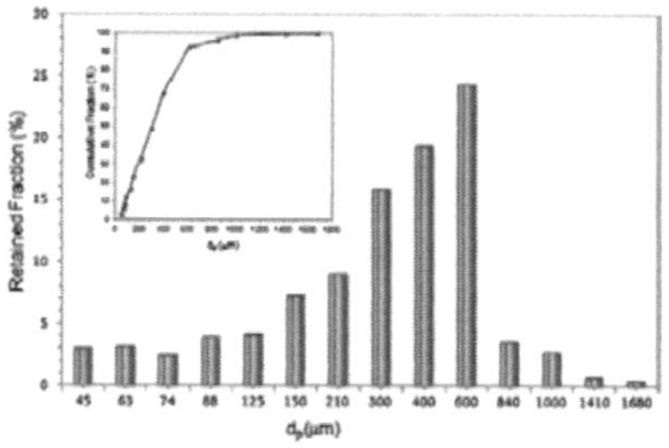

Size distribution for Douglas fir wood chips ground in a hammer mill at 1.6 mm screen size, *Source: Tannous 2013*

7

Low-Order Intra-Particle (LOIP) Model

Particle Heat-Up

- Most important temperature gradients are along narrowest dimension
- Approximate heat-up as 1-D conduction with average properties and simple B.C.

$$\rho C_p \frac{\partial T}{\partial t} = \frac{1}{r^b}\frac{\partial}{\partial r}\left(k\, r^b \frac{\partial T}{\partial r}\right) + g$$

$$k \left.\frac{\partial T}{\partial r}\right|_{r=R} = h\,(T_\infty - T_R)$$

$$\left.\frac{\partial T}{\partial r}\right|_{r=0} = 0$$

where ρ = density (kg/m³)
C_p = heat capacity (J / kg·K)
k = thermal conductivity (W / m·K)
T = temperature (K)
T_∞ = ambient temperature (K)
T_R = surface temperature (K)
r = radius (m)
b = shape factor of 0=slab, 1=cylinder, 2=sphere
g = heat generation (W/m³)
h = heat transfer coefficient (W / m²·K)

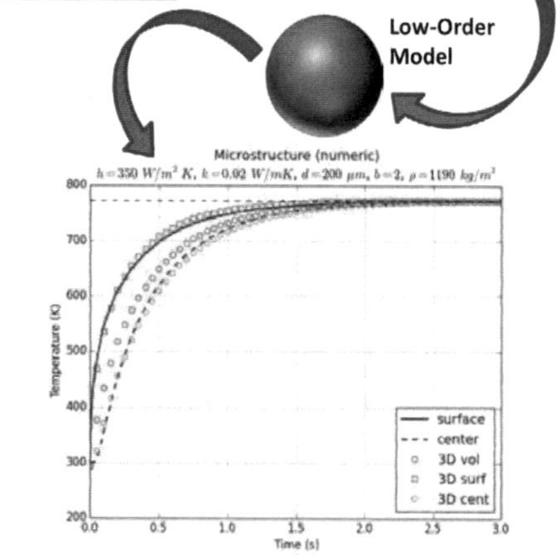

Temperature profile of 3-D microstructure model (NREL) and low-order intra-particle model. Heat transfer via surface convection and conduction only.

8

LOIP kinetics are based on available data

Available kinetics are typically very simplified

- Primary and secondary reactions produce gas, tar (condensable liquid or bio-oil), and char
- Non-condensable gases
  - light gases such as H_2, CO, CO_2, H_2O, CH_4, etc.
- Condensable volatiles
  - heavy organics and inorganics
  - vapors and aerosols react within and/or outside particle
- Char as solid residue
- Typically 1st order Arrhenius

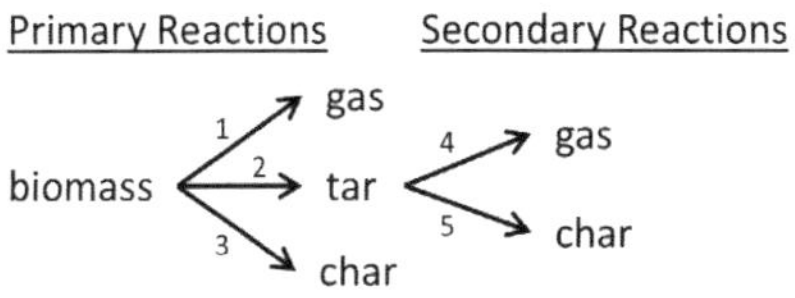

Chan1985 and Blasi1993 kinetic parameters

Reaction	A (1/s)	E (kJ/mol)
1	1.3e8	140
2	2e8	133
3	1.08e7	121
4	4.28e6	108
5	1e6	108

$$K = Ae^{-E/(RT)}$$

$$r = K \cdot \rho_i$$

$$\rho_f = \rho_i + r \cdot \Delta t$$

where K = rate constant (1/s)
A = pre-factor (1/s)
E = activation energy (kJ/mol)
R = universal gas constant (kJ / mol·K)
r = reaction rate (ρ/s)
ρ = concentration (kg/m³)
Δt = time step (s)

LOIP model is configured to allow multiple user specified kinetics and internal transport schemes to account for future improvements in kinetic information.

9

Other LOIP Assumptions

- Particle does not shrink or expand during pyrolysis
- Thermal properties (c_p, k) vary with temperature
- Catalytic effects of ash are neglected
- Gases immediately leave the particle
- Effects of mass transport are not accounted for

Will be implemented in future models.

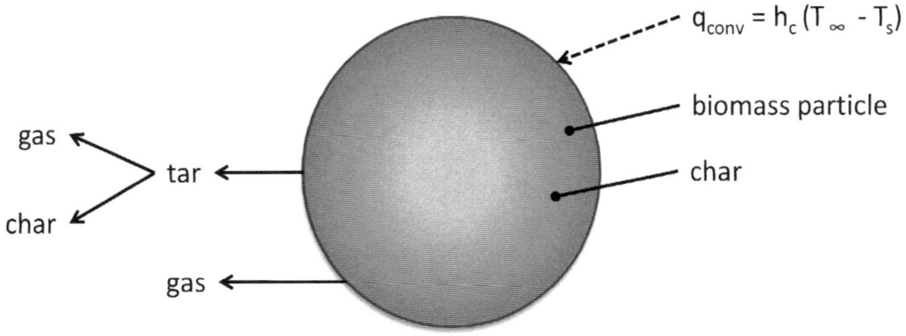

T_∞ = 425-600°C (698-873K)

10

LOIP appears to reasonably approximate 3D heat-up

Low-order vs 3-D Microstructure Model

- Lumped capacitance method agrees with volume avg. 3-D temperature profile for skeletal density of 1190 kg/m^3
- 1-D numerical approach provides good agreement using minimum particle width, spherical shape factor, and k_{eff} value same as N_2 gas
- Lumped method approximated by V/A_s , 1-D transient heat conduction approximated via smallest particle dimension

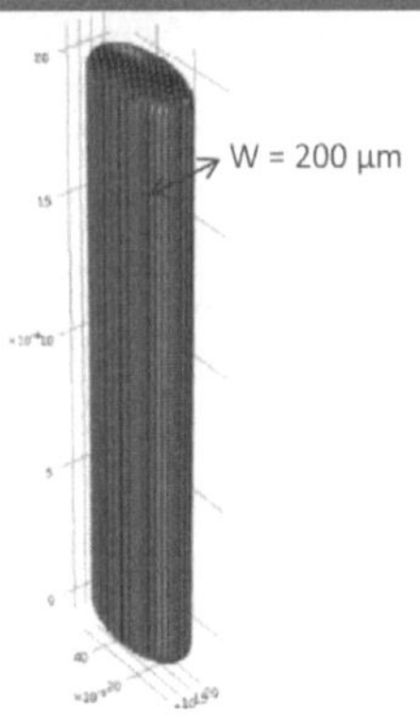

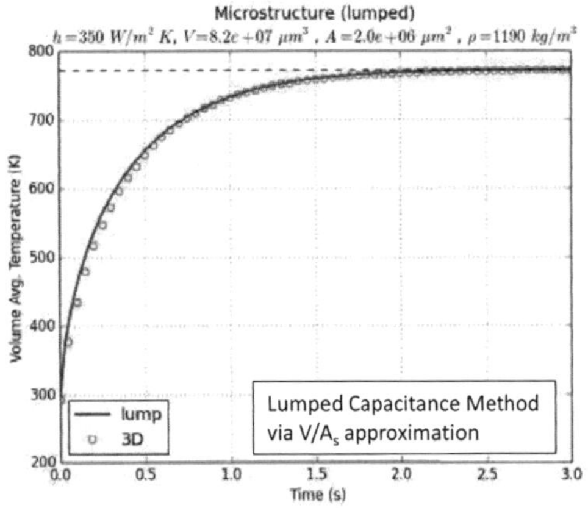

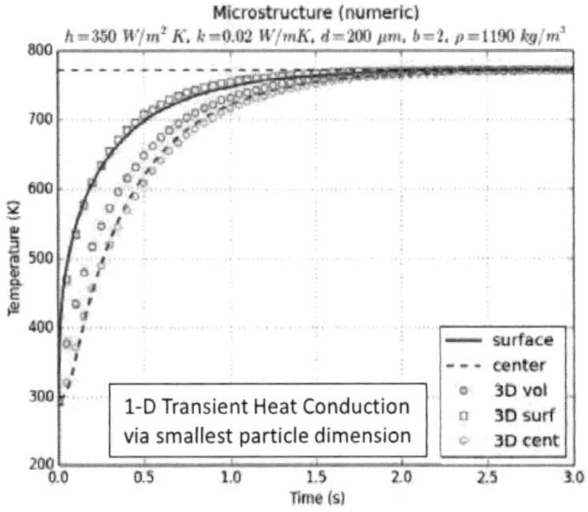

11

LOIP heat transfer and kinetics reveal important trends

Trends from 1-D transient heat conduction model.

Best Case Scenarios: Offers a liquid yield of ≈64% for a dry 2.0 mm spherical particle. As particle size increases, time to reach peak liquid yield increases.

Primary Reactions

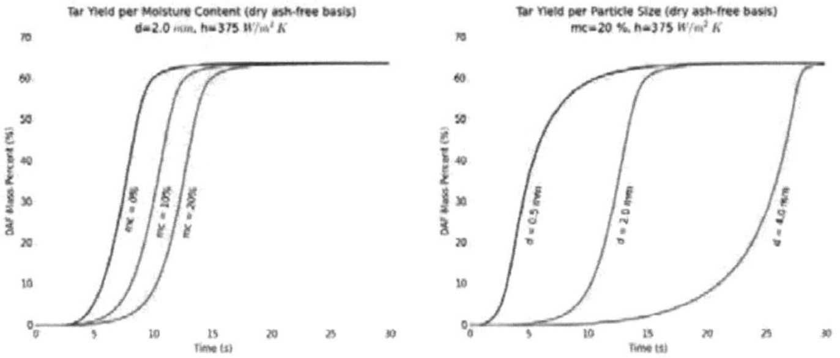

Tar yield (aka liquid or bio-oil yield) at different moisture contents (Left) and particle sizes (Right). Represents optimal liquid yield (no secondary reactions) where vapor and gas immediately leave particle.

Worst Case Scenarios: Drastically decreases max liquid yield to ≈27% for a dry 2.0 mm spherical particle. As particle size increases, time to reach peak liquid increases while the max yield decreases.

Primary+Secondary Reactions

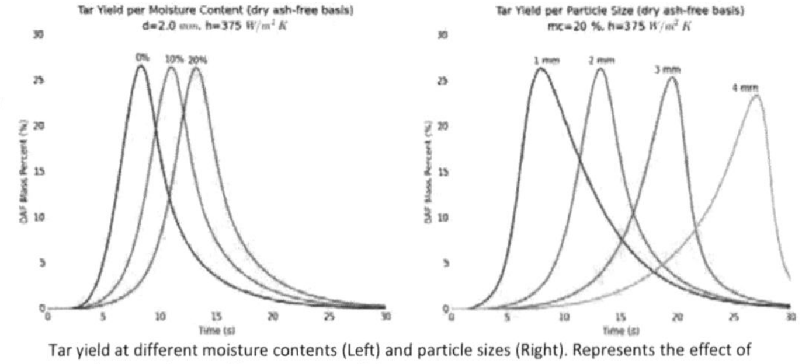

Tar yield at different moisture contents (Left) and particle sizes (Right). Represents the effect of secondary reactions cracking tar to gas and char.

12

Particle and Gas Mixing Challenges

Turbulent mixing of biomass particles in bubbling beds is highly complex.
[Xiong 2013, Papadikis 2010, Radmanesh 2005]

Mixing depends on particle properties (size, density) and local flow conditions and involves recirculation and segregation.
[Di Blasi 2008, Cui 2007, Wang 2005, Kunii 1991]

Internal temperature and concentration gradients can be significant as reactor scale increases.

Detailed CFD simulations of mixing require high computational overhead and time, which increase with reactor scale.
[Mellin 2014, Papadikis 2009]

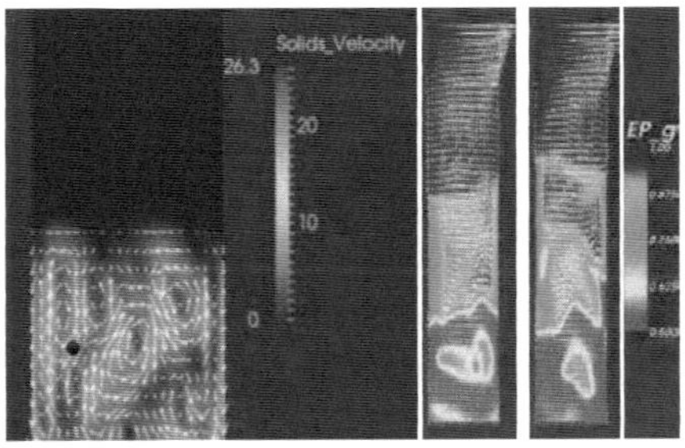

MFIX simulation of pyrolysis bubbling bed - (Left) solids velocity cm/s and (Right) particle void fraction.

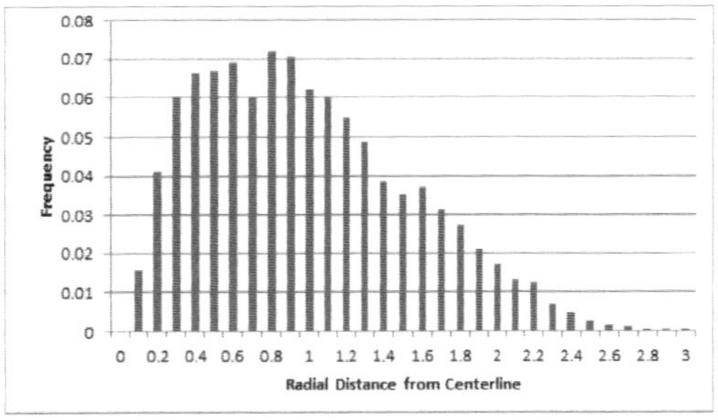

Radial distribution for entire bed of particles in an experimental fluidized bed, *Source: Separation Design Group.*

13

Low-Order Particle Mixing (LOPM) Model

Langevin Model

$$m\frac{d^2x}{dt^2} = -\lambda\frac{dx}{dt} + \eta(t)$$

where x = distance, t = time, m = particle mass,
λ = friction coefficient,
η(t) = stochastic perturbations

Paul Langevin
(1872-1946)

- Originally proposed by Paul Langevin to describe Brownian motion
  Langevin, Paul. "Sur la théorie du mouvement brownien." CR Acad. Sci. Paris 146.530-533 (1908)

- We propose a modified version of this model for biomass particles in bubbling beds for the **vertical direction**

$$m\,\frac{d^2z}{dt^2} = -\lambda\,\frac{dz}{dt} + f_s + f_{t_z}$$

where f_s = mean segregation force in axial direction, f_{tz} = turbulence force in axial direction

- A similar force balance can be written for **horizontal particle position** except that we assume no time-average drag or gravitational forces

$$m\,\frac{d^2h}{dt^2} = -\lambda\,\frac{dh}{dt} + f_c + f_{t_h}$$

where f_c = horizontal force via convection toward wall, f_{th} = turbulence force in horizontal direction

14

The LOPM is formulated as a random walk

- Approximating derivatives over discrete time intervals and combining and rearranging terms for **vertical motion**

$$z(t+1) = a \cdot z(t) + b \cdot z(t-1) + c + s_z(t)$$

where $z(t)$ = axial position at time t, $s_z(t)$ = turbulence term, Gaussian distributed
a, b, c = empirical parameters that reflect time average forces

- a, b, and c can be estimated with experimental particle position time series
- Stochastic inputs (s_z, s_r, s_h) can be estimated from stepwise prediction errors
- For **horizontal motion**

$$h(t+1) = a \cdot h(t) + b \cdot h(t-1) + s_r(t)\, g(z) + s_h(t)$$

where $h(t)$ = lateral position at time t, $s_r(t)$ = probability particles entrained in sideways convective flow
$g(z)$ = amplitude of local convective flows with height, $s_h(t)$ = small scale turbulence, Gaussian distribution
a, b = empirical parameters that reflect time average forces

- In vectorized form, the random walk can be applied to thousands of particles simultaneously and adjusted at each time step for changing particle properties

See Ind. Eng. Chem. Res., 2014, 53 (41), pp 15836–15844 for more details.

15

Experimental validation of LOPM predictions

- Simulated biomass (tracer) particles are constructed by inserting tiny neodymium magnets into balsa wood cylinders (typically > 1 mm diameter, 0.4-1 g/cc)
- Bed particles (207 μm glass, 2.5 g/cc) are fluidized with ambient air ($1.0 \leq U/U_{mf} \leq 5.0$)
- Single tracer particles are injected in bed at specific fluidization conditions and tracked
- Special algorithms de-convolute signals to give 3-D particle trajectory
- Experimental facility at Separation Design Group Lab, designed and operated by Jack Halow

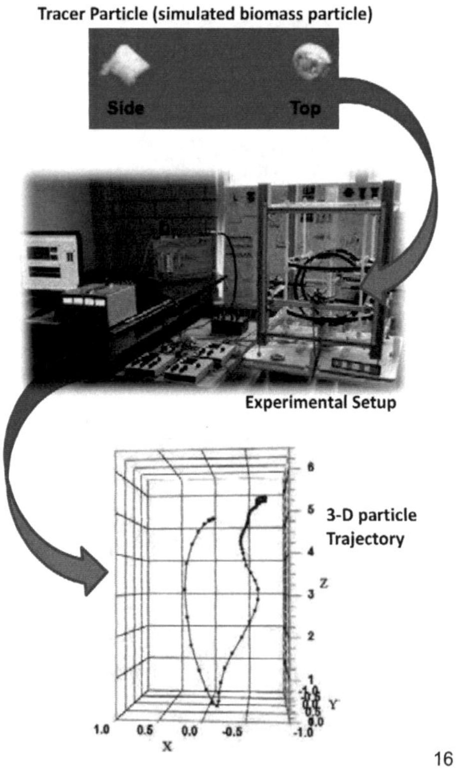

16

Both LOPM & experiments yield particle RTDs

Trajectories map of 3-D time-average mixing.

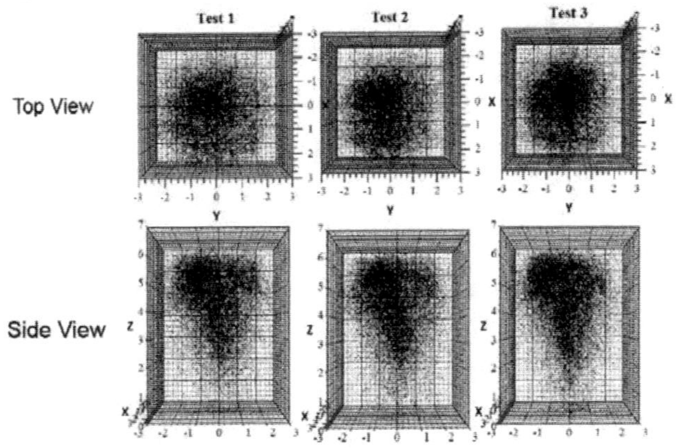

Vertical mixing profiles follow Weibull statistics.

$$f(z) = \frac{k}{\lambda}\left(\frac{z}{\lambda}\right)^{k-1} e^{-(z/\lambda)^k}$$

$$C(z) = 1 - e^{-(z/\lambda)^k}$$

where f(z) = probability density function
C(z) = cumulative distribution function
k = shape parameter
λ = scale parameter
z = axial position

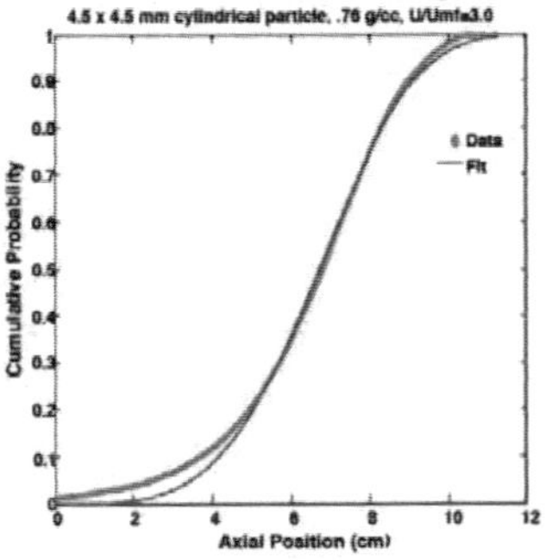

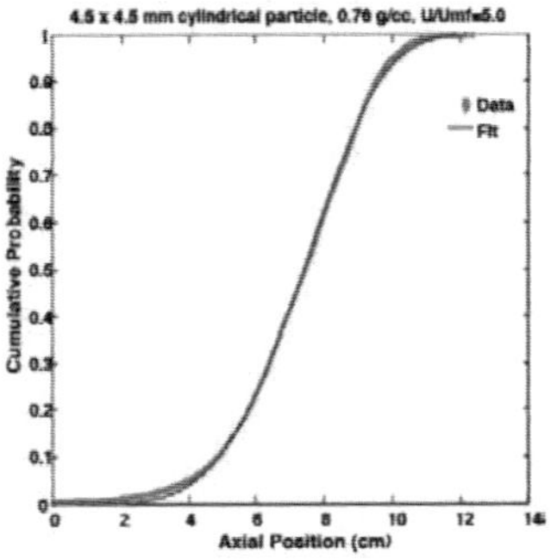

17

Integrated Pyrolysis Reactor Model

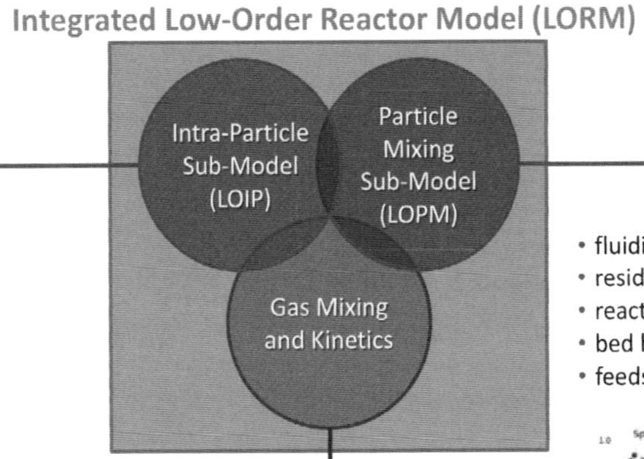

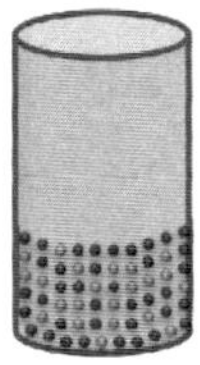

- heat transfer
- mass transport
- thermal properties
- physical characteristics
- moisture content

- fluidization scheme
- residence times
- reactor height / diameter
- bed height
- feedstock injection

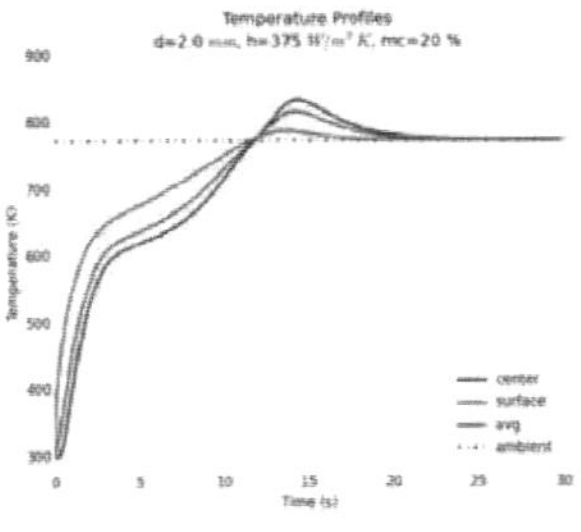

- chemical species
- devolatilization rate
- liquid yield
- secondary reactions

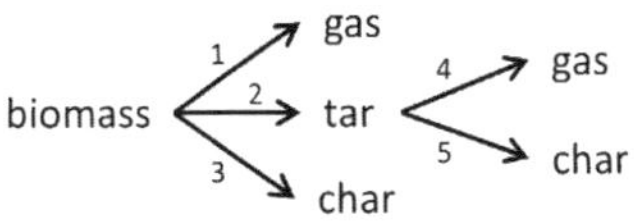

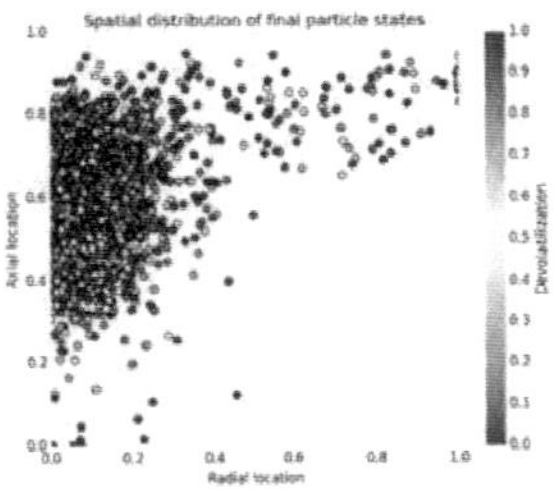

- LOIP sub-model tracks gas release and evolution of multiple particle sizes
- LOPM sub-model mixes particles via random walk with changing parameters
- Released gases undergo further reaction and mixing before exiting reactor

18

Results of LOIP and LOPM studies to date

- Complex 3D intra-particle heat transfer appears to be reasonably well approximated by simplified 1D model with appropriate thermal property assumptions
- Available simple pyrolysis kinetics appear to yield realistic trends in liquid yield with particle size and moisture, but more detailed kinetics are needed to account for liquid product composition
- Biomass particle RTDs generated by simple random walks appear to be consistent with magnetic tracer measurements and exhibit Weibull statistics
- More experimental/CFD data are needed to correlate the random walk parameters with particle properties and fluidization conditions

19

Next Steps...

- Add more detailed intra-particle mass transfer and kinetics (Ranzi2008, 2013) to the LOIP
- Add improved estimates for the impact of particle properties on the LOPM random walk parameters (based on CFD and/or experiments)
- Develop sensitivity trends (for operating, feedstock, and scale parameters) in raw bio-oil yield and quality from combined LOIP+LOPM simulations
- Compare above sensitivity predictions with available experimental lab and pilot data
- Conduct/select specific experiments needed to confirm prediction accuracy of models and better inform scale-up

20

Next Steps…

- Model gas RTD as combination of PFRs and CSTRs based on CFD and experimental measurements

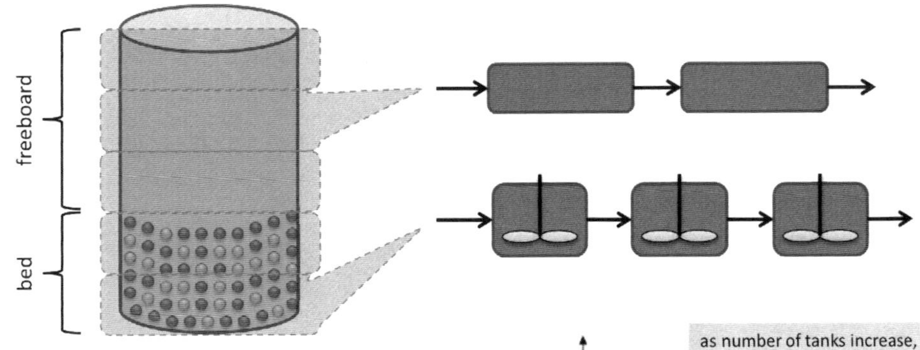

$$E(t) = \frac{t^{\,n-1}}{(n-1)!\,\tau_i^n}\, e^{-t/\tau_i}$$

where $E(t)$ *= residence time distribution function*
t *= time*
n *= reactors*
τ_i *= residence time in one of the reactors*

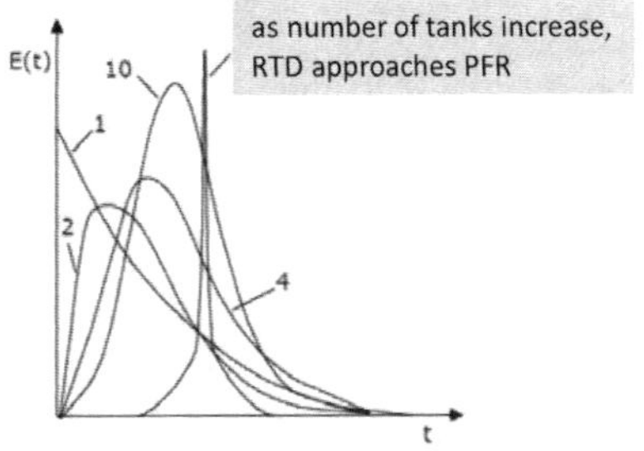

RTD for 1, 2, 4, 10 tanks and a PFR, *Source: Wittrup 2007*

21

Acknowledgements

Prasad Gupte: DOE Project Manager
U.S. Department of Energy, Bioenergy Technologies Office

Emilio Ramirez, Sreekanth Pannala, Charles Finney, Stuart Daw
ORNL CPC Team Members

<u>Collaborators</u>
Peter Ciesielski, *National Renewable Energy Lab*
Victor Walker, Tyler Westover, *Idaho National Lab*
Jack Halow, *Separation Design Group*

22

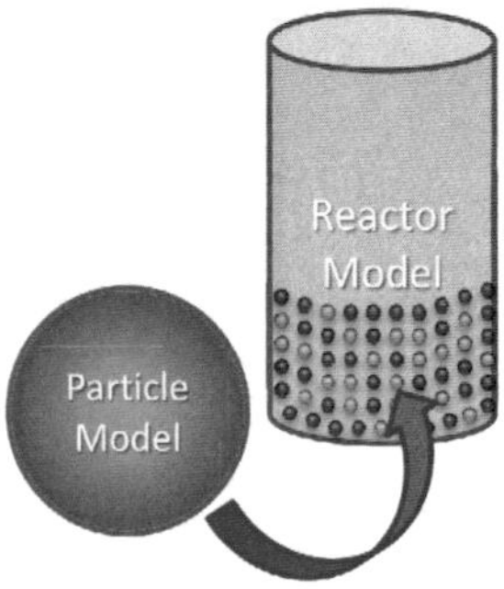

Publications on the LOIP and LOPM sub-models are in process and supplemental information will be available on the CPC website.

Gavin Wiggins: wigginsg@ornl.gov

Computational Pyrolysis Consortium

CPC website: http://energy.ornl.gov/cpc

GitHub

Python code: https://github.com/pyrolysis

23

Backup Slides

24

Transient Heat Transfer Equations

Low-order Particle Models

- 1-D transient heat conduction

$$\rho\, c\frac{\partial T}{\partial t} = \frac{1}{r^b}\frac{\partial}{\partial r}\left(k\, r^b\frac{\partial T}{\partial r}\right) + g$$

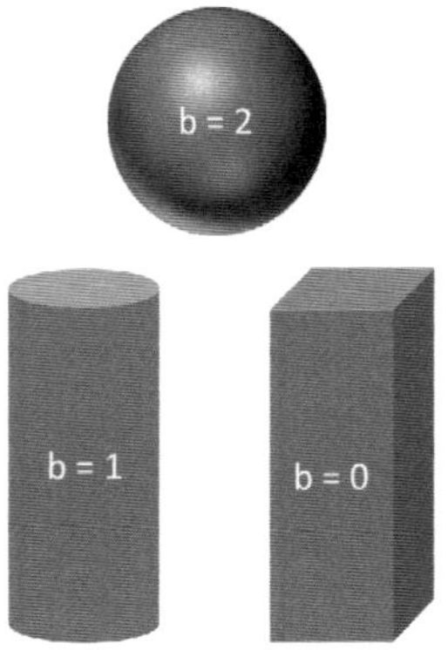

- Lumped capacitance method (Bi < 0.1)

$$\frac{\theta}{\theta_i} = \frac{T - T_\infty}{T_i - T_\infty} = \exp(-Bi \cdot Fo)$$

- Improved lumped capacitance method (Bi <= 20)

$$Bi = \frac{h\, L_c}{k}$$

$$\theta_p = \exp\left(-\frac{1}{\frac{m+1}{m+3}\, Bi + 1}\, Bi\, Fo\right)$$

$$Fo = \frac{\alpha\, t}{{L_c}^2}$$

Source: Keshavarz 2006

25

Parameters and Assumptions

3-D Models	Low-order Models
volume and density of sphere and shell are equal, $\rho = 540\ kg/m^3$	
microstructure density represented as cell wall density	
heat transfer via conduction and convection at surface (no kinetic reactions)	
k = 0.12 W/m·K	shell diameter as average of L+W
C_p = (103.1+3.86*T) J/kg·K	k = 0.12 (wood) - 0.02 (N_2 gas) W/m·K
identical thermal capacity	C_p = (103.1+3.86*T) J/kg·K
spatial distribution of mass	

V_A (solid) < V_B

mass = $V_A\, \rho_A = V_B\, \rho_B$

$\rho_A = V_B\, \rho_B / V_A$

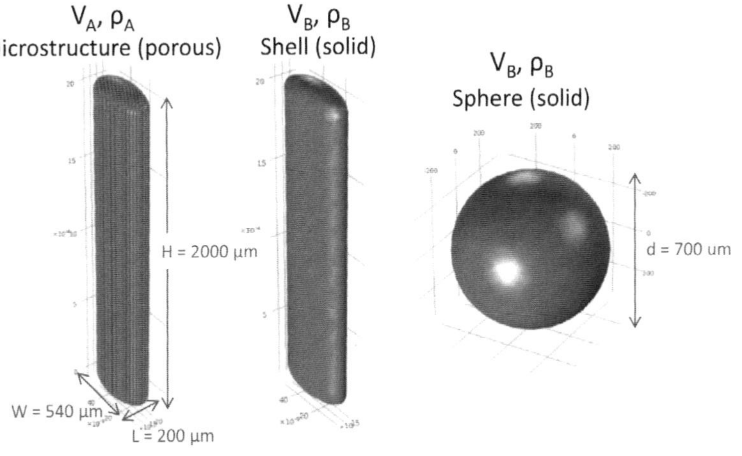

26

Particle parameters affect pyrolysis yields thru the reaction steps

Even the simplest proposed kinetic schemes involve both parallel and sequential competing reactions.

Primary Reactions   Secondary Reactions

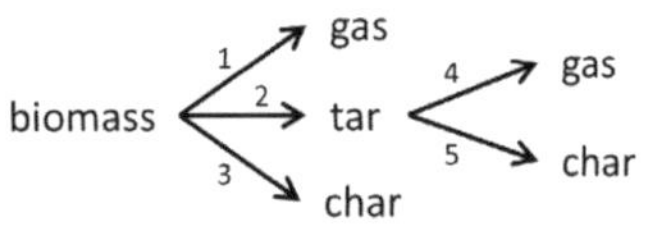

Reaction	A (1/s)	E (kJ/mol)
1	1.3e8	140
2	2e8	133
3	1.08e7	121
4	4.28e6	108
5	1e6	108

Best Case Scenarios: Liquid yield is high if the optimal tar formation temperature is reached quickly and the tar products escape quickly from the particle and encounter no further char or catalysts in the reactor to drive the secondary reactions.

Worst Case Scenarios: Liquid yield is low if tar formation is slow (vs. the primary gas and char reactions) and/or the initial tar has opportunities to crack within or outside the particle. Particle mass transport and reactor mixing determine the importance of the secondary reactions.

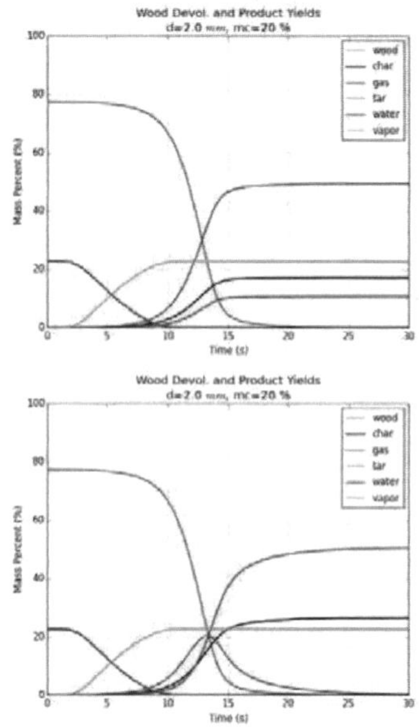

Wood devolatilization and product yields for 2 mm particle at 20% moisture content.

(Top) Optimal conditions within particle
(Bottom) Worst-case scenario

27

Experimental Setup

- Probes aligned North, South, East, West
- Helmholtz coils modify earth's magnetic field in bed
- Non-metallic bed and supports
- 100 Hz sampling rate, 5 minute runs (30,000 data points)
- Porous plate distributor

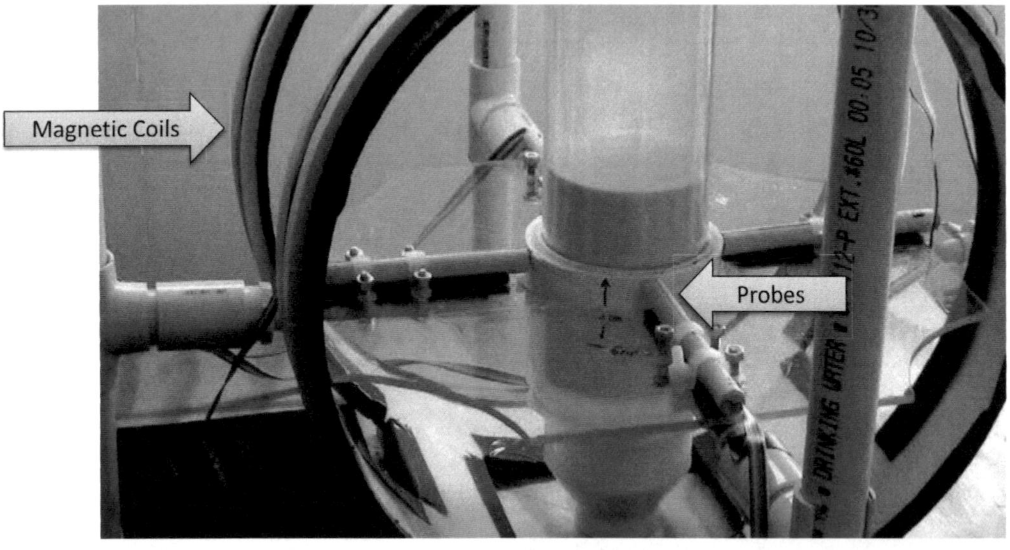

28

Experimental Setup

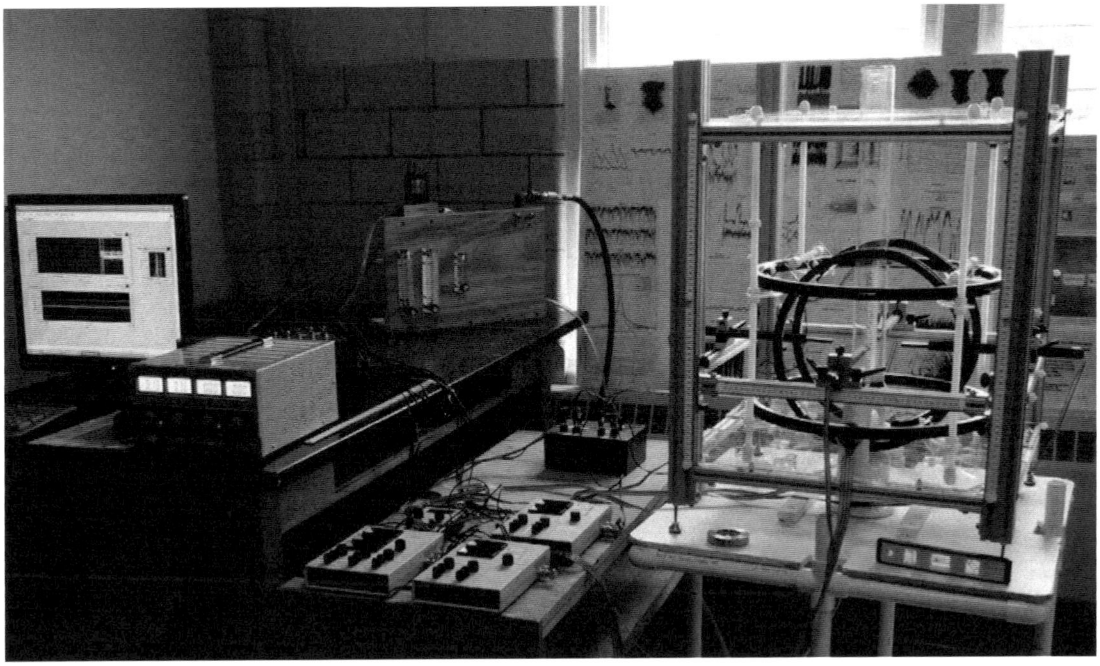

29

Experimental Setup

30

Time series data reveal particle motion dynamics

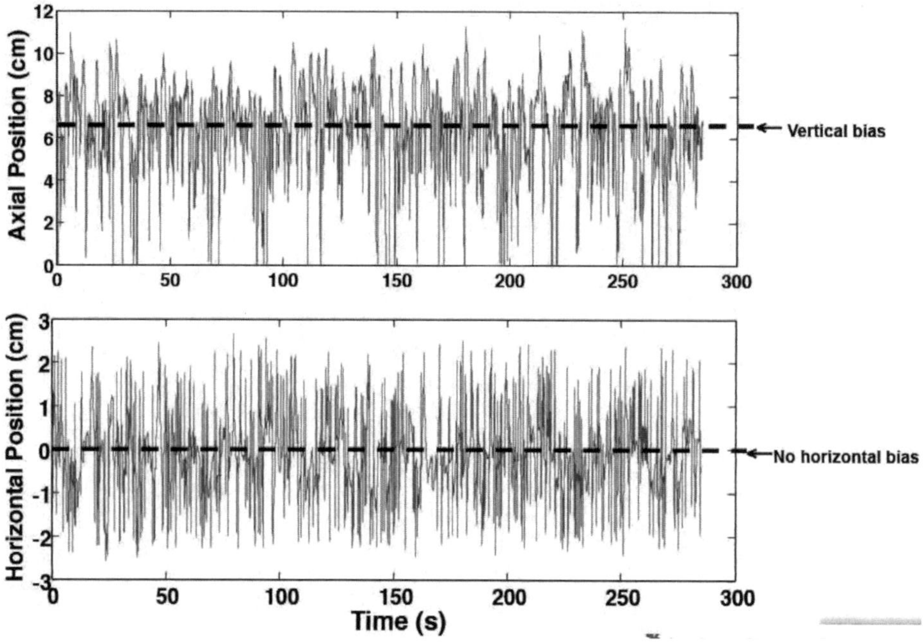

31

Next Steps...

- Couple the Ranzi kinetic scheme to the heat transfer model to estimate chemical species from biomass pyrolysis

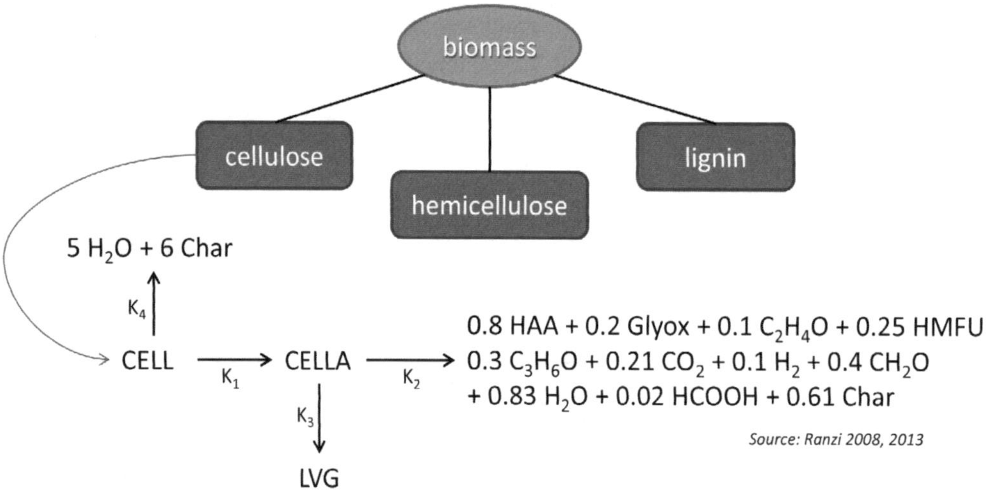

Source: Ranzi 2008, 2013

- Develop an appropriate definition for the “diameter” or “size” of a wood particle that is used for pyrolysis modeling

32

Computational Pyrolysis Consortium (CPC)

Members of the Consortium

Oak Ridge National Laboratory

- Coordination of the CPC team with industry advisors and university partners
- CFD of biomass pyrolysis reactors
- Low-order models for pyrolysis and upgrading reactors
- Multi-stage model integration

NREL – National Renewable Energy Laboratory

- Micro-to-pilot-scale reactor data
- Biomass particle-scale reaction and transport models
- Catalytic vapor-phase kinetic models
- CFD of vapor-phase catalytic upgrading
- Identification of critical TEA inputs

Pacific Northwest National Laboratory

- Hydro-treating and aqueous upgrading catalyst data
- Non-polar and polar liquid phase catalytic kinetic models
- Integrated liquid catalytic reactor models
- Identification of critical TEA inputs

INL – Idaho National Laboratory

- Biomass feedstock characterization tools and data
- Model component and data sharing/archiving mechanism

Argonne National Laboratory

- Vapor-phase catalytic molecular energetics
- Fundamental bio-oil vapor thermodynamic properties
- Identification of potential catalysts for vapor-phase upgrading

Assess the commercial feasibility of advanced catalytic technologies for producing infrastructure compatible transportation fuels from biomass-derived pyrolysis oils.

33

Computational analysis of biomass particles in a bubbling bed

Emilio Ramirez[1], Stuart Daw[1], Janine Galvin[2], Jack Halow[3], and Sreekanth Pannala[1], and Charles Finney[1]

[1] Oak Ridge National Laboratory

[2] National Energy Technology Laboratory

[3] Separation Design

Abstract

A leading candidate process for converting biomass to liquid fuels involves fast pyrolysis of the biomass in a fluidized bubbling bed reactor of inert or catalytic particles. As the biomass heats through its contact with hot gases and solids, a complex series of thermally driven reactions releases volatile constituents, ash, and char. The pyrolysis reaction rates and product gas composition, which determine the yield of fuel-compatible molecules, are highly dependent on the heat and mass transfer occurring both within the biomass particles themselves and also at the interface between the particles and their surrounding environment as they transit through the reactor.

In this study, we simulate the effects of biomass particle properties and fluidization state on the biomass particle mixing and residence time distribution in a pyrolysis reactor. For our simulations, we utilized MFIX, an open source, computational software package supported by DOE, which uses a continuum (two-fluid) approach for efficient modeling of the hydrodynamics (Garg et al. 2012, DOI: 10.1016/j.powtec.2011.09.019). We apply a novel discrete element method (DEM) algorithm to simulate the mixing behavior of individual biomass particles in a background of bed particles. The output of our simulations includes detailed statistics about the expected biomass particle trajectories.

For our initial validations, we compare our simulation results with measurements of simulated biomass particle mixing in an experimental cold-flow fluidized bed of Geldart B particles designed to replicate many of the features of a bubbling bed pyrolysis reactor. The laboratory measurements included detailed 3-D spatial trajectories of magnetically tagged simulated biomass particles in the laboratory bubbling bed (Halow et al., 2012, DOI: 10.1021/ie301517w). Comparisons are made between the simulations and experimental measurements for a range of fluidizing gas flows and biomass particle size and density. In addition, we report results from a broader parametric study with our modified MFIX model to identify general sensitivities and mixing trends in bubbling bed biomass pyrolysis reactors.

377349 Hydrodynamics Study on Wind-Box and Gasifier Zone of Dual Fluidized-Beds By CFD Simulation

Abstract:

The hydrodynamic characteristics in a wind-box and gasifier zone of dual fluidized-bed (DFB) are investigated by a computational fluid dynamics (CFD) model to find a stable flow rate zone of fluidizing gas in a pilot-scale biomass gasification process. The gas and granular phases are treated by an Eulerian-Eulerian two-fluid model, coupling with the k-epsilon realizable turbulent model. The random motion of solid particles in the granular phase is represented with the help of the kinetic theory of granular flow (KTGF). A time-step adaptation procedure is implemented to provide the solution stability in the wide range of turbulence and to achieve the numerical solution efficiency for the huge number of mesh elements. The three-dimensional CFD model is used to identify hydrodynamics of bubbling fluidization using air as the fluidization agent and sand as the heat carrier particles in the pilot-scale DFB at an operating temperature of 800 °C. The simulation results provide detailed information of gas and solid flow regimes, pressure drops, temperature and particles distribution, and bubble behaviors. Effects of different air flow rates on hydrodynamics of the bubbling gasification zone are examined and the stable air flow rate zone is proposed.

392454 Advanced Biorefinery Towards a Sustainable Bioeconomy: Value-Added Biobased Materials

Abstract:

In both the biochemical and thermo-chemical bioconversion process we generate a considerable amount of downstream products. The value-added uses of co-products and bio-products from biofuel industries are termed as "Advanced Biorefinery" in this presentation. Dry milling corn ethanol industries produce distillers' grains and carbon dioxide whereas lingo-cellulosic ethanol industries produce lignin and carbon dioxide as their downstream products. The thermo-chemical conversion of biomass produces bio-oil, syngas and bio-chars. The value-added biobased materials from co-products/byproducts will not only help in enhancing the profitable return of the related renewable biofuel industries but also can help in substituting certain petroleum-based products in the manufacturing sectors. These new biobased materials can be cost-competitive with added advantages of eco-friendliness. The recent developments of such biobased materials for potential uses in consumer products, packaging and interior automotive parts will be discussed.

Acknowledgement: The financial support from the Natural Sciences and Engineering Research Council (NSERC) Canada for the Discovery Grant; NSERC NCE AUTO21; the Ontario Research Fund (ORF) Research Excellence (RE) Round-4 from the Ontario Ministry of Economic Development and Innovation (MEDI); the Ontario Ministry of Agriculture and Food (OMAF) and Ministry of Rural Affairs (MRA) - University of Guelph Bioeconomy-industrial uses research program; and OMAF-MRA New Directions Research Program to carry out this research is gratefully acknowledged

368896 Selective Conversion of Biorefinery Lignin to Dicarboxylic Acids

Abstract:

Lignin is a ubiquitous component in almost all plant biomass. Large quantities of industrial lignin are already produced annually as a waste product of the pulp and paper industry, where the vast majority is burned as a low cost fuel to provide energy for the papermaking process. The emerging biomass refinery industry will further introduce an enormous amount of lignin. Thus, there is an urgent need to develop technologies that can create new applications for biorefinery lignin. Lignin is the largest source of renewable material with an aromatic skeleton.

Majority of previous in area have been focused on depolymerizing lignin to low molecular weight phenolic compounds (LMWPC). However, one potential group of chemicals has been overlooked among oxidative conversion products: dicarboxylic acids (DCA). DCA such as muconic, maleic, and succinic acids are important and highly valuable industrial chemicals and intermediates used in many industries including the biopolymer, pharmaceutical, and food additives industries. Current commercial dicarboxylic acids are all produced from petroleum based feedstock. Developing a green route to producing DCA from renewable biomass lignin will be of a prime interest to both chemical and biomass conversion industry. Oxidative aromatic ring cleavage to yield muconic acid and its derivatives is a well-recognized reaction. Although DCA and their derivatives have been detected in the reaction mixture from oxidative delignification of wood pulp, there has not been successful attempt to produce DCA from biomass lignin. Discovering a selective route to convert lignin to DCA will pave a new avenue towards lignin utilization.

This paper demonstrates a novel route to creating dicarboxylic acids from biorefinery lignin via ring cleavage by chalcopyrite catalyzed oxidation in a highly selective process. We reported a method of selective production of DCA via chalcopyrite ($CuFeS_2$) catalysed oxidation of biorefinery lignins in the presence of hydrogen peroxide (H_2O_2) under mild reaction conditions. A clear two stage reaction mechanisms is proposed, including depolymerization and oxidative aromatic ring cleavage, as explaination of the products profiles through lignin and model compounds study. Up to 95% selectivity of stable, saturated dicarboxylic acids was obtained from several types of biorefinery lignin and model compounds under mild, environmentally friendly reaction conditions.

Making activated carbon from low/un-hydrolyzed biomass residue

Chen Li and Sandeep Kumar

Department of Civil & Environmental Engineering, Old Dominion University, Norfolk, VA, USA

Corresponding author: skumar@odu.edu

Bioethanol has become one of the most promising biofuels for short to medium alternative to fossil transportation fuels. During production of bioethanol, a large amount of un-hydrolyzed solids (UHS) is generated. The UHS resulting from enzymatic hydrolysis during cellulosic ethanol production is a wet and carbon-rich residue. It mainly contains lignin and unconverted sugars depending upon the types of biomass pretreatment used in the earlier stages. Several tons of UHS are expected to be generated during cellulosic ethanol/butanol production from lignocellulosic biomass. The efficient utilization of this residue will have a positive impact in making the biofuels cost-competitive. In recent few decades, adsorption has become a well-established technique to remove pollutants from water and air. Activated carbon is becoming the prevailing adsorbent for the purification of water with low pollutant concentration due to their advantage in high internal surface areas and active surfaces. However, the major drawbacks of commercially available carbons are expensive, so more efficient and low-cost ones now are needed for water industry[1].

Recently, there have been lots of efforts on research in developing recycling processes to produce activated carbon from biomass waste, for instance, thermal conversion of pyrolysis and gasification processes [2]. There are already many kinds of biomass waste such as apricot stone shells, bean pods, and corn stover are being used to produce activated carbon, [3]. Under certain process conditions several activated carbons were produced from agricultural residues having BET surface areas ranging between 250 and 2410 m^2/g and pore volumes of 0.022 and 91.4 cm^3/g [4].

In this study, the UHS generated after enzymatic scarification of corn stover is used as a feedstock for activated carbon production. Hydrothermal processing of biomass under subcritical water condition is another promising green technology to extract useful sugar from biomass. Large amount of solid residue is recovered after the hydrothermal treatment. The other feedstock including corn stover and pinewood are first subjected to hydrothermal hydrolysis to extract sugars and the residues are used for activated carbon production. Our objective is to hydrolyze maximum sugar in aqueous phase and use the low sugars containing solid residue for activated carbon production. In this study, solid residues after hydrothermal hydrolysis are named low-hydrolyzed biomass residue (LHR).

Two-step experiment is designed. The first step is focused on maximizing hydrothermal sugar extraction result and producing LHR. During the hydrothermal hydrolysis, reaction time between 3 mins to 8 mins, reaction temperature between 180°C to 260 ^{O}C, and sulfuric acid concentration (upto 1 v/v%) are studied. The second step is focused on making activated carbon form UHS and LHR. The high temperature chemical activation is our primary method to make activated carbon. The effect of reaction time (30 mins to 120 mins) and temperature (750°C to 800 ^{O}C) on the BET surface area are studied. Different chemicals such as $ZnCl_2$, H_3PO_4 and K_2CO_3 are tested. The quality evaluation matrix of activated carbon included the analysis of BET surface areas, pore volumes, iodine value, and burn off percentage. One of the experiments result showed that the surface area of activated carbon from LHR was 658m^2/g and the pore volume was 0.13cc/g. The objective of the study is to optimize the process for high quality activated carbon production from LHR and UHS.

References

1. Budinova, T., et al., *Biomass waste-derived activated carbon for the removal of arsenic and manganese ions from aqueous solutions.* Applied Surface Science, 2009. **255**(8): p. 4650-4657.
2. Daifullah, A. and B. Girgis, *Removal of some substituted phenols by activated carbon obtained from agricultural waste.* Water research, 1998. **32**(4): p. 1169-1177.
3. Tay, T., S. Ucar, and S. Karagöz, *Preparation and characterization of activated carbon from waste biomass.* Journal of Hazardous Materials, 2009. **165**(1): p. 481-485.
4. Ioannidou, O. and A. Zabaniotou, *Agricultural residues as precursors for activated carbon production—a review.* Renewable and Sustainable Energy Reviews, 2007. **11**(9): p. 1966-2005.

392455 Reactive Blending of Protein Rich Meals and Biodegradable Polymers for Green Packaging

Abstract:

Plastics are highly valued materials because of their low cost and extraordinary versatility and they constitute the second largest application of petroleum next to energy. Efforts to replace commodity plastics based packaging materials with biodegradable counterparts are usually hampered by strong completion with cheaper petroleum-based polymers and limited performance. Blending of relatively expensive biodegradable polymers such as poly (butylene succinate) (PBS), poly (butylene adipate co-terephthalate) (PBAT), and polycaprolactone (PCL) with low value and cheap agricultural feedstock can reduce the cost of the resulting materials while maintaining their biodegradability. For example, the use of less expensive protein rich meals such as soybean meal, canola meal, corn meal etc. obtained from the food and biodiesel industry as co-reactants of biodegradable polymers in green packaging applications is expected to reduce the cost competitiveness barrier. Reactive extrusion processing could be one of the potential techniques with a lot of promise for future development of biodegradable plastics. This is because, it is a continuous and cost effective polymer processing techniques that allows melt blending and various other chemical reactions including (co) polymerization, compatibilizing, grafting, branching and functionalization without the need for solvents. The direct use of relatively cheap soy protein meal in the reactive blending, and study of compatibilization chemistry to develop a cost competitive green packaging films are the main objectives of this research. The soy meal underwent through three types of fermentations to de-structurize the ingredients of the soy meal including simple sugars and cellulose that hinders film making behavior. Post-fermentation, the protein and pectin concentrations in the soy meal have significantly increased. Also, plastic films made out of the blending of the fermented soymeal and PBAT exhibited a significant performance improvement.

Acknowledgement: The financial supports from 1) the Ontario Ministry of Agriculture and Food (OMAF) and Ministry of Rural Affairs (MRA) - University of Guelph Bioeconomy-industrial uses research program; 2) OMAF-MRA New Directions Research Program; and 3) the Ontario Research Fund (ORF) Research Excellence (RE) Round-4 from the Ontario Ministry of Economic Development and Innovation (MEDI) and Natural Sciences and Engineering Research Council (NSERC) Discovery grant individual (to Mohanty) to carry out this research are gratefully acknowledged.

387114 Separate Hydrolysis and Fermentation of Untreated and Pretreated Miscanthus Cake to Produce Ethanol

Abstract:

The use of lignocellulosic biomass to produce biofuel will add value to land and reduce emissions of greenhouse gases by replacing petroleum products. Valuable co-products derived from fractionation of miscanthus give the resulting fibrous fraction an economic advantage as a feedtock for ethanol production. Freshly harvestedmiscanthus from the North Carolina A and T farm was dewatered using press, whereby miscanthus is separated into a fiber-rich cake and a nutrient-rich juice. The production of cellulosic ethanol from miscanthus fibers with or without alkali(NaOH) pretreatment were investigated by this work using separate hydrolysis and fermentation (SHF). The objective was to compare the effect of alkali pretreatment on the miscanthus samples with regards to ethanol yield.

The pretreatment of miscanthus with NaOH was carried out to assess the effect of various NaOH concentrations on the enzymatic digestibility of the wet miscanthus cake. The concentrations of NaOH investigated were 0, 1, 4 and 7%, Enzymatic hydrolysis of the miscanthus cake was carried out using a cocktail of enzymes comprising of cellulase (NS 50013), β-glucosidase (NS 50010) and hemicellulose (NS 22002) at a temperature of 50oC for 96 hours. This was followed by fermentation using Escherichia coli (E. coli) at a temperature of 32oC, also for 96 hours.

The hydrolysis products investigated using High Performance Liquid Chromatography (HPLC) included glucose, xylose, arabinose and uronic acids. Samples were further analyzed for fermentation products including ethanol, glucose, xylose, arabinose and acetic acid. The results show a steady increase in fermentable sugars from the enzymatic hydrolysis over the time period, and also an increase in ethanol produced from the pretreated miscanthus as compared to the untreated miscanthus with the overall effectiveness of the alkali pretreatment being a function of the NaOH concentration. This suggests effective delignification leading to enhanced ethanol production.

Chemical activation of biomass-derived hydrochar for environmental applications: Influence of processing conditions during hydrothermal treatment

Akshay Jain[a], *Rajasekhar Balasubramanian*[b] & *M.P. Srinivasan*[*a]

[a] Department of Chemical and Biomolecular Engineering, National University of Singapore, Singapore 117585, Singapore.

[b] Department of Civil and Environmental Engineering, National University of Singapore, Singapore 117585, Singapore

Abstract

Coconut shell-derived hydrochar has been used as precursors to produce mesoporous carbonaceous adsorbents with high mesopore surface area (570- 1150 m^2/g). The textural properties of the activated carbon can be easily tuned by modifying the concentration of reactants (coconut shell, $ZnCl_2$ and water). The tuning of the stoichiometry of reactants governed the formation of oxygenated functional groups on hydrochar and $ZnCl_2$ solubility and thus its availability at the time of activation. Up to 36% increase in mesopore area and 18% increase in BET area could be achieved by the use of hydrothermal pre-treatment compared to conventional soaking. A combination of hydrothermal pre- treatment using H_2O_2 followed by the hydrothermal treatment using $ZnCl_2$ at different concentrations resulted in a significant improvement in the mesopore surface area (from 570 to 1150 m^2/g) and BET surface area (from 1720 to 2440 m^2/g) of activated carbon. Increase in mesopore area is validated by the increased adsorption capacity of organic dyes.

Use of simple and resource-/energy- efficient pre-treatment processes will be the key to designing high-surface area mesoporous adsorbents for environmental applications such as wastewater treatment and gas separation.

The use of Amazonian biomass for small biorefineries serving small communities in the rainforest

Angela A. González, Juan C. Higuita, Paola A. Moreno, Carlos A. Cardona
agonzalezvil@unal.edu.co , jchiguitav@unal.edu.co, palmorenonu@unal.edu.co , ccardonaal@unal.edu.co

Instituto de Biotecnología y Agroindustria
Universidad Nacional de Colombia
Manizales - Colombia

14AIChE
Annual Meeting, Atlanta, GA

Amazon Region

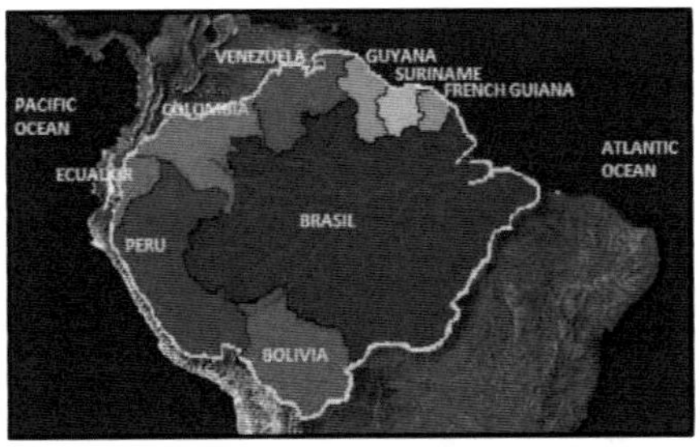

Fig 1. Amazon Region. Adapted from WWF

Table 1. The Amazon Region area per country

Country	Area (km^2)	%
Bolivia	724.000	9,76
Brazil	5,034,740	67,91
Colombia	477,274	6,43
Ecuador	115,613	1,56
Guyana	214,960	2,89
Perú	651,440	8,79
Suriname	142,800	1,93
Venezuela	53,000	0,73
Total	7,413,827	100

Source from: Palacios (2011)

- 56% of the world rainforest.
- 28 million people (4.5 inhabitants / km^2).
- 385 indigenous groups.

It represents the 40% South America territory.

Research Group in Chemical, Catalytic and Biotechnological Processes
Universidad Nacional de Colombia Manizales Campus 2

Colombian Amazon Region

Economy: Natural resources exploitation.

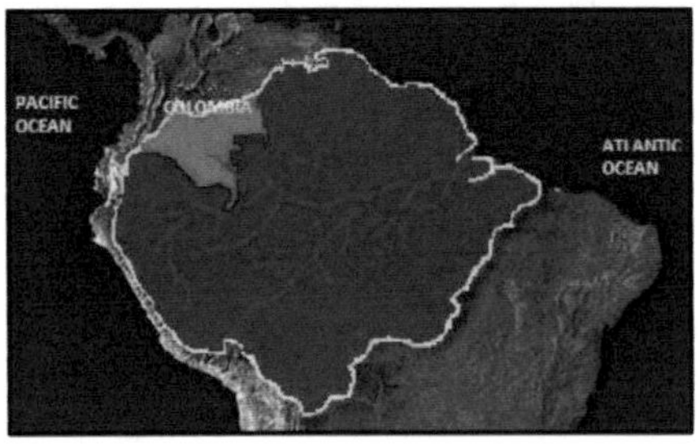

Fig 2. Colombian Amazon Region. Adapted from WWF

Colombian Amazon Region: 483.164 km² (41% of the Country area)

By 2011 the region recorded poverty rates of 35,6%

Source from: https://www.dane.gov.co/files/investigaciones/condiciones_vida/pobreza/cp_pobreza_2011.pdf

The economy of Colombian Amazon Region

Natural resources exploitation.

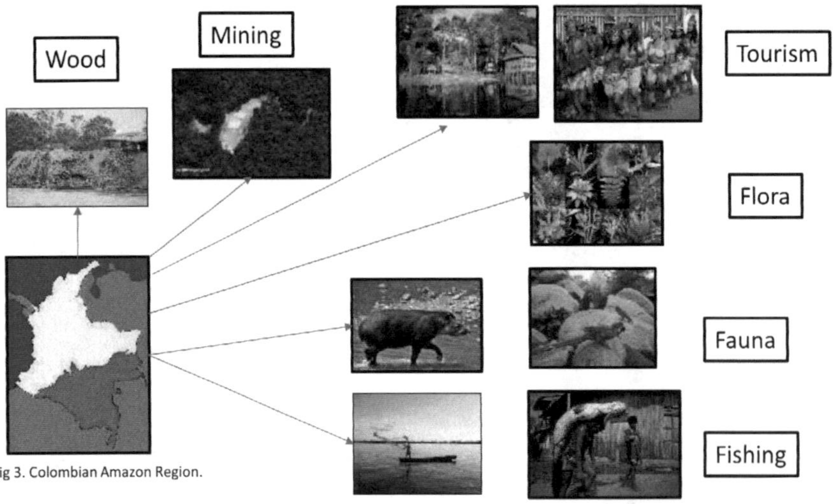

Fig 3. Colombian Amazon Region.

Source from: www.lugaresde viaje.com, http://geografiadelamazonas.blogspot.com.es /, www.keywordpictures.com, www.andigena.org, http://es.mongabay.com/, www.imeditores.com

Entering to the Colombian Amazon

Fig 3. Amazonas - Colombia.

- Amazonas: Provence at south of Colombia.
- Connectivity limited.
- Air and water communication. NO roads.

Area: 109,665 km^2 (equivalent to Cuba)
9,6% of the Continental Territory
PIB participation: 0,1%
90,000 inhabitants (0,8 inhabitant/km^2)

Entering to the Colombian Amazon

Fig 3. Amazonas - Colombia.

- Potential of production:
  - Agro: fruits, medicinal plants.
  - Agroindustry: cacao, juices, pulps, vitamins...
  - Wood: cedar, mahogany.
  - Services: tourism, environmental.

The study case

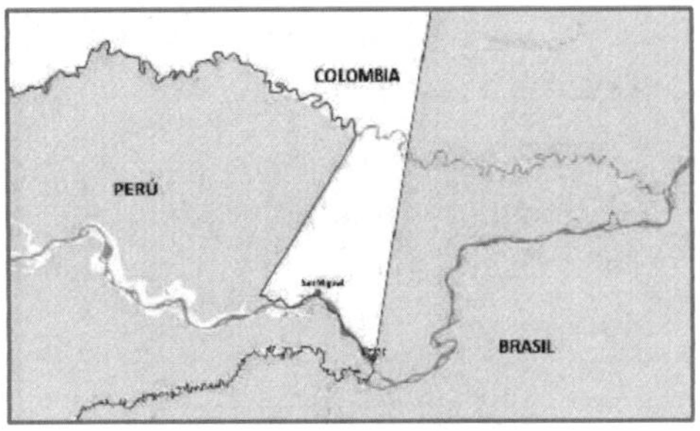

Fig 4. San Miguel Amazonas Colombia

- San Miguel.
- Small community on the Amazon River two hours from Leticia.
- Fluvial communication.
- 300 inhabitants.

Local initiative of production

Cacaoteros del Amazonas is an association of ten natives families (Tikuna) that they have been producing cacao since 4 years ago.

Currently they market handmade cocoa.

Cocoa production

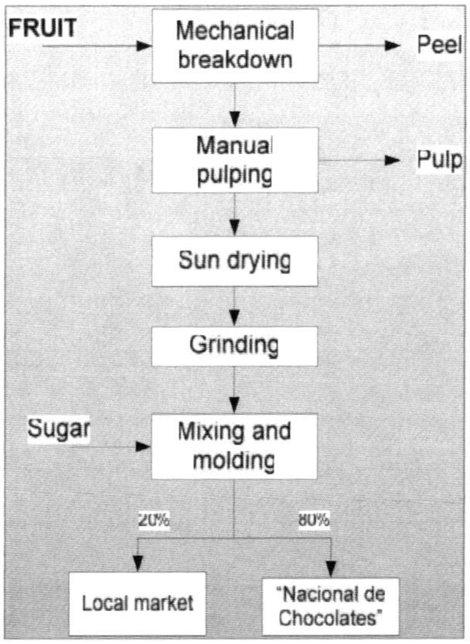

Sale Price:
2,81 US/kg product cocoa

Fig 6. Production process flow sheet according to the Cacaoteros

"Amazonian Cocoa, an economic alternative for Amazonas"

Research Group in Chemical, Catalytic and Biotechnological Processes
Universidad Nacional de Colombia Manizales Campus

9

Objective

To evaluate technical, economic and environmental the process for obtaining products with add-value from *Theobroma cacao* in the Colombian Amazon context.

Research Group in Chemical, Catalytic and Biotechnological Processes
Universidad Nacional de Colombia Manizales Campus

10

Cocoa – *Theobroma cacao*

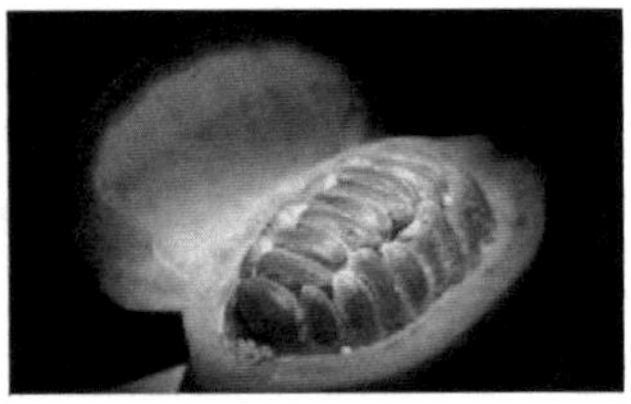

Table 2. Cocoa composition used for simulation process

Feature	Peel[1]	Pulp[2]	Seed[3]
Percentage	75,0	4,0	21,0
Moisture (%)	8,5	80,0	4,6
Ash (%)	6,7	0,5	3,6
Protein (%)	8,6	0,1	20,1
Lipid (%)	1,5	N.A.	54,5
Carbohydrates (%)	32,3	13,0	11,4
Lignin (%)	21,4	N.A.	5,8
Fiber (%)	36,6	1,1	N.A.

1 Vriesmann et. al. (2011).
2 Liendo et. al. (1997).
3 Sánchez V.A. (2007) .

Research Group in Chemical, Catalytic and Biotechnological Processes
Universidad Nacional de Colombia Manizales Campus
11

Methodology

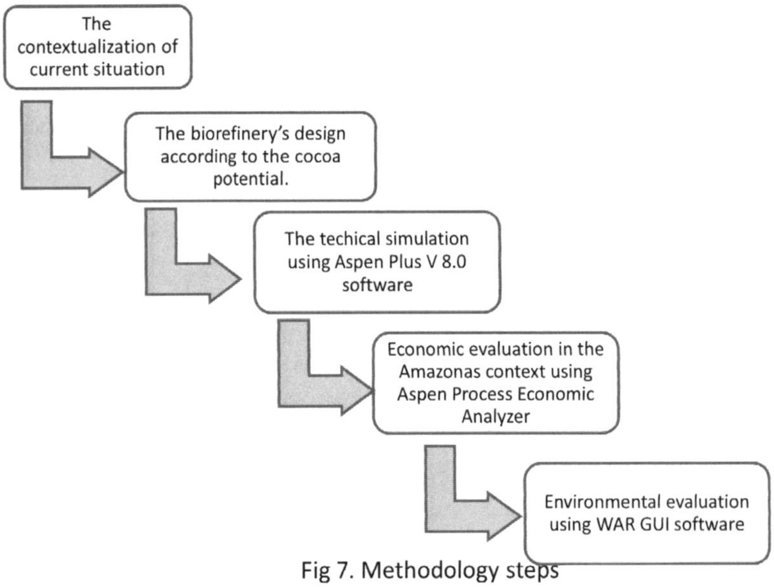

Fig 7. Methodology steps

Research Group in Chemical, Catalytic and Biotechnological Processes
Universidad Nacional de Colombia Manizales Campus
12

Results

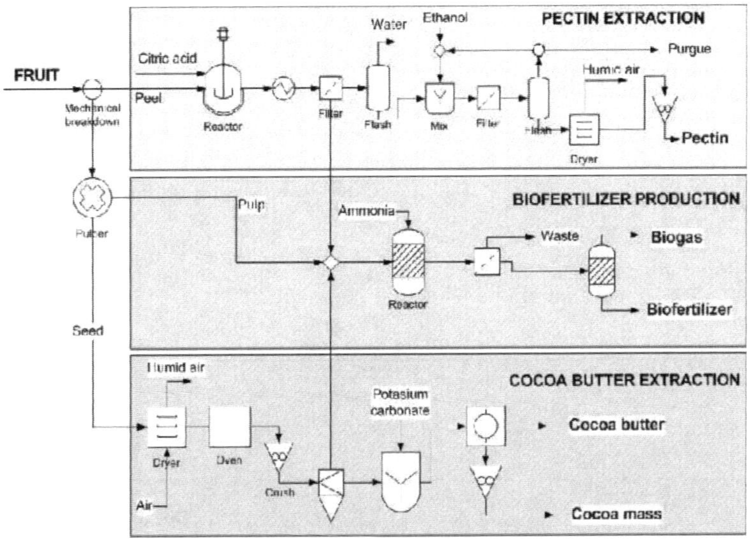

Fig. 8 Biorefinery flow sheet

Scenarios

According to agricultural production capacity of San Miguel.

Table 3. Proposed scenarios

SCENARIO	RAW MATERIAL (TON/DAY)
1	0,5
2	1
3	2

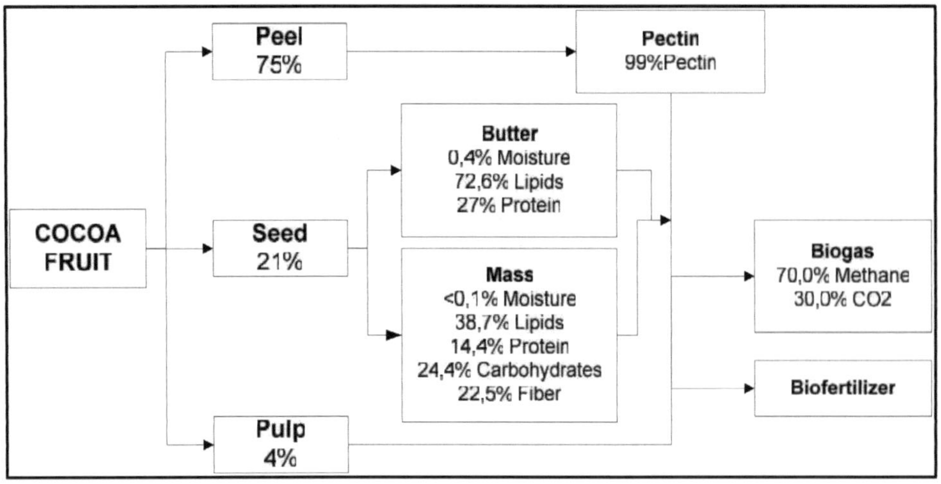

Fig. 9 Quality of cocoa biorefinery products

Research Group in Chemical, Catalytic and Biotechnological Processes
Universidad Nacional de Colombia Manizales Campus 15

Distribution cost

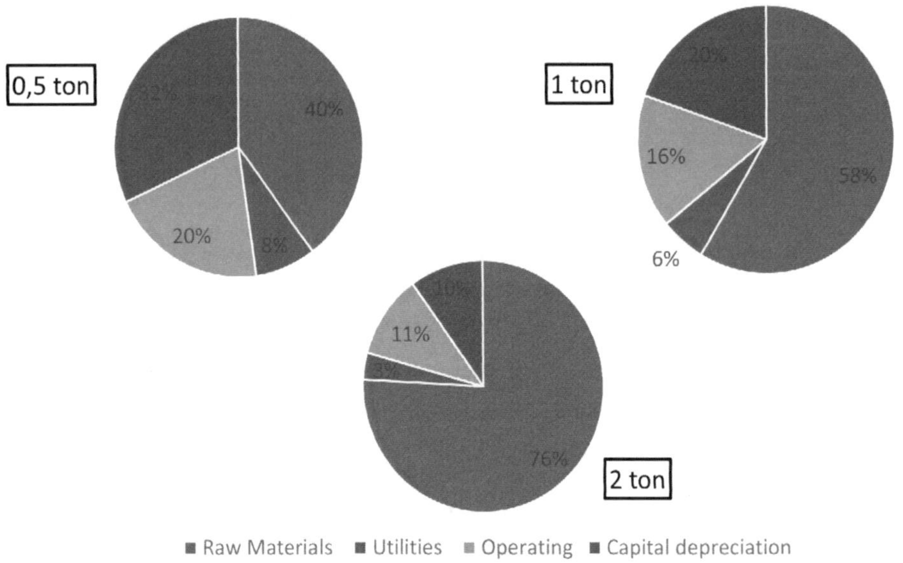

Fig. 10 Total distribution cost of Cacao Biorefinery

Research Group in Chemical, Catalytic and Biotechnological Processes
Universidad Nacional de Colombia Manizales Campus 16

Distribution cost

Table 4. Cost of production per product

SCENARIO	PECTIN	BUTTER	MASS	BIOGAS	BIOFERTILIZER
	USD/kg	USD/kg	USD/kg	USD/kg	USD/kg
0,5 ton	37,47	0,48	0,60	13,10	2,12
1 ton	28,77	0,20	0,25	9,99	1,62
2 ton	27,87	0,15	0,19	8,42	1,37

Research Group in Chemical, Catalytic and Biotechnological Processes
Universidad Nacional de Colombia Manizales Campus 17

Environmental analysis

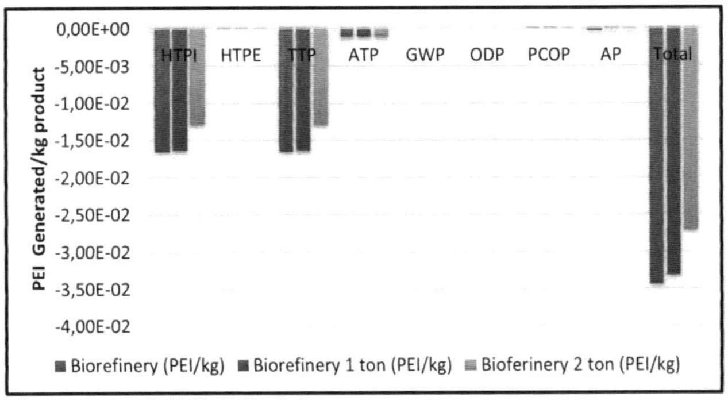

Fig. 11 Biorefinery Potential Environmental Impact generated

HTPI: Human Toxicity Potential by Ingestion.
HTPE: Human Toxicity Potential by Exposure.
TTP: Terrestrial Toxicity Potential.
ATP: Aquatic Toxicity Potential.
GWP: Global Warming Potential.
ODP: Ozone Depletion Potential.
PCOP: Smog Formation Potential.
AP: Acidification Potential.

Research Group in Chemical, Catalytic and Biotechnological Processes
Universidad Nacional de Colombia Manizales Campus 18

Conclusions

According this study it was possible conclude:

- The production under biorefinery concept is a possibility to take in count in the Amazonas Provence.

- The simulation methodology can help in the research of productive Amazon initiatives because it reduces the pressure over ecosystems and allows to evaluate previously the technical, economic and environmental feasibility of the project.

Research Group in Chemical, Catalytic and Biotechnological Processes
Universidad Nacional de Colombia Manizales Campus 19

REFERENCES

- Palacios, Pablo. 2011. "Devenir Amazónico. La Transformación Del Paisaje Amazónico." In *Amazonia Colombiana: Imaginarios Y Realidades. Cátedra Jorge Eliécer Gaitán*, edited by Catalina Pérez Niño Juan Alvaro Echeverri. Bogotá.
- Vriesmann L.L., de Mello A., R.D, de Oliveira P. C.L. (2011) *Cacao pod husk (Theobroma cacao L.): Composition and hot-water-soluble pectins.* Industrial Crops and Products 34 1173-1181.
- Barazarte, H., Sangronis, E., Unai, E. (2008) *La cáscara de cacao (Theobroma cacao L.): una posible fuente comercial.* Archivos latinoamericanos de nutrición.Vol. 58 N°1.
- Redgwell, R., Trovato, V., Merinat, S., Curti, D., Hediger, S., Manez, A., (2003) *Dietary fibre in cocoa Shell: characterisation of component polysaccharides.* Food chemistry 81, 103-112.
- Liendo R., Padilla F.C., Quintana A., (1997) *Characterization of cocoa butter extracted from Criollo cultivars of Theobroma cacao L.* Food Research International. Vol. 30. No. 9 727 – 731.
- Sánchez C., V.A., (2007) *Caracterización organoléptica del cacao (Theobroma cacao L.), para la selección de árboles con perfiles de sabor de interés comercial.* Tesis de grado. Universidad Técnica estatal de Quevedo.
- Cardona, C.A. & Cerón, I.X., (2012) Conceptual design of sustainable biorefineries - Metabolites and bioenergy production in the Amazoon. The World Congress on Industrial Biotechnology and Bioprocessing, Orlando, Florida.
- Moncada, J., El-Halwagi, M. M., & Cardona, C. A. (2013). *Bioresource Technology, 135*(0), 533-543.
- Rincón, L., Becerra, L., Moncada, J., & Cardona, C. (2014). *Waste and Biomass Valorization, 5*(2), 189-198.
- Rincón, L. E., Hernández, V., & Cardona, C. A. (2014). *Process Biochemistry, 49*(3), 474-489.

Research Group in Chemical, Catalytic and Biotechnological Processes
Universidad Nacional de Colombia Manizales Campus 20

Corresponding Autor: Carlos Ariel Cardona Alzate.
Km 9 Vía al Aeropuerto La Nubia, Manizales, Caldas
tel. +57 68879400 ext 55354
e-mail: ccardonaal@unal.edu.co
Universidad Nacional de Colombia. Sede Manizales

Research Group in Chemical, Catalytic and Biotechnological Processes
Universidad Nacional de Colombia Manizales Campus

One-pot conversion of cellobiose into polyols and the possible reaction pathways over extremely low phosphotungstic acid with Ru/C catalyst

Kai Zhang[a], *Shubin Wu*[a]*, *Xiaohong Zhang*[b]
[a]*State Key Laboratory of Pulp and Paper Engineering, South China University of Technology, Guangzhou, China*
[b] *Guangzhou Institute of Energy Conversion,Chinese Academy of Science,Guangzhou, China*
Corresponding author. Tel./fax: +86 2022236808
E-mail address: shubinwu@scut.edu.cn

Abstract

Cellobiose is a glucose dimer connected by a glycosidic bond and represents the basic repeating unit of cellulose, and it could be efficiently converted into polyols in the presence of hydrogen and precious-metal catalyst of Ru/C over extremely low phosphotungstic acid which is environmentally friendly and reduces the requirement for equipments. Then, processing parameters and the reaction performance were studied in detail, the yields of ethylene glycol (EG) and 1,2-propylene glycol (1,2-PG) were obtained up to 33.4% and 12.2% respectively at 240 ℃ for 1 h, which had a definite guidance function for catalytic conversion of cellulose.Besides, the catalytic process using cellobiose was studied to propose the possible reaction pathways of producing polyols .

Keywords Catalytic conversion; Cellobiose; Polyols; UPLC-MS; Pathway;

Introduction

In order to alleviate the shortage of fossil fuels and reduce the CO_2 emissions, the renewable biomass is a good choice for producing liquid fuels and value-added chemicals[1-2]. As the most abundant source of biomass, cellulose is currently thought as a promising alternative for fossil fuels because of its direct conversion into useful organic compounds without imposing a negative impact on food supplies [3-4]. However, unlike other water soluble polysaccharides, the low reactivity of cellulose has prevented its use in chemical industry except for the paper manufacturing[5]. In order to utilize cellulose, the key factor is to destory the glycosidic bonds connecting glucose residues to promote its dissolution in solvents. Fortunately, Luo et al[6] presented a green approach to efficient conversion of cellulose into polyols in hot water (about 473K) which could genenrate H^+ ions to catalyze the hydrolysis reactions of cellulose. Besides, solid catalyst is proved to be an environmentally benign and milder reaction in hydrolytic conversion of cellulose[7-8]. Then, Zhang Tao and coworkers[9] reported a one-pot conversion of cellulose to EG with 61% yield over a nickel-promoted tungsten carbide catalyst in hot water, which has attracted much attention from both academic and industrial societies since 2008.

It is reported that EG is an important chemical product, widely used in the synthesis of cosmetics, polyester fibers, antifreeze, etc, and its global consumption reached more than 21 million tons in 2011[10]. With the increasing demand for EG in the plastics and automotive industry, its direct production from cellulose will creat a renewable way for reducing dependence on petroleum. Therefore, more and more attentions are going to be paid to this field, meanwhile, a lot of research results have been attained. For example, combined with Ru-based or Ni-based catalyst, tungsten-based catalysts such as W, WO_3, H_2WO_4, heteropoly acids (PW, SiW), etc were found to have a higher selectivity to ethylene glycol (EG) on conversion of cellulose or glucose in presence of hydrogen[11-13]. Thereinto, Zhao Guanhong and coworkers[11] found that in combination with 4wt% Ru/AC, tungstophosphoric acid(PTA) showed a better performance on selectively producing 1,2-PG(12.7%) in addition to EG(30.1%) from the conversion of glucose. 1,2-PG is also an important bulk chemical widely used as materials of plasticizer, surfactant, resin, emulsifier and anti-emulsifier, heat carrier, antifreeze and so on[14].

However, the concentration of PTA containing 1000ppm tungsten is high in the above reaction, which leads to problems such as acid corrosion and hard disposal, and cellulose or cellobiose could used to directly produce polyols in presence of PTA over Ru/C catalyst[15]. Therefore, low concentration of PTA should be used in the reaction and the performance on selectively producing polyols from the conversion of cellulose or cellobiose should be studied.

Herein, we report a one-pot catalytic process using cellobiose as material over 5wt% Ru/C catalyst in combination with extremely low PTA which is environmentally friendly and reduces the requirement for equipments. Then, processing parameters and the reaction performance are studied in detail to have a definite guidance function for catalytic conversion cellulose. While, cellulose is a polymer with a complex molecular structure, using simple model molecules of cellulose to study the reaction process presents a promising way, and cellobiose is a glucose dimer connected by a glycosidic bond and represents the basic repeating unit of cellulose[16-17]. Therefore, according to the results from catalytic process using cellobiose and research progress from others, the possible reaction pathways of producing polyols are proposed.

Experimental Section

Materials:

Ru/C (precious-metal loading:5 wt.%) was purchased from Aladdin reagent limited company. Tungstophosphoric acid and cellobiose were obtained from Tianjin Kemiou Chemical Reagent Co. Ltd and Shanghai bioscience and technology limited company respectively.

Catalytic Reaction:

Catalytic conversion of cellobiose was typically carried out in a high pressure stainless-steel autoclave (Parr Instrument Co, 5500, 100mL). Catalysts and tungstophosphoric acid solution, cellobiose was preloaded in the reactor. Then, the reactor was filled with H_2 after expelling air, the reactor was heated to the reaction temperature from room temperature(RT), and the reaction temperature was kept a certain times. After reaction, the reactor was cooled down to RT rapidly with ice water. The liquid products were collected by filtration with 0.22 μm membrane, then they were stored in the refrigerator.

Product Analysis:

After the reaction, the liquid products were analyzed with liquid chromatography (Waters 2695 HPLC equipped with Aminex HPX-87H 300×7.8mm column, refractive index and ultraviolet detector). Besides, in order to further verify the reaction pathways of producing polyols, Ultra Performance Liquid Chromatography tandem mass spectrometry(Agilent 1290 UPLC equipped with mass spectrometry 6540 and zorbax carbohydrate 4.6× 150mm column) was used.

Yield(%)=(weight of polyol in the products)/(weight of cellulose put into the reactor)×100%

Results and Discussion

1 The effect of processing parameters on the conversion of cellobiose

The effect of processing parameters such as H_2 pressure, temperature, time, and etc on the yield of polyols were investigated over extremely low phosphotungstic acid with Ru/C catalyst. When a parameter was studied, other processing parameters were kept consistent. Thereinto, the processing parameters were chosen as follows: the reaction temperature at 513K and reaction time for 1 h, 5Mpa of the H_2 pressure and 1% content of cellobiose in the reaction system, 0.1g of 5wt% Ru/C, 600rpm and 30mL extremely low phosphotungstic acid in water(0.01wt%).

1.1 The effect of H_2 pressure on the yield of polyols from cellobiose conversion

Fig 1 shows that with the increase of the H_2 pressure from 3Mpa to 5Mpa, the yield of EG reaches to the highest value of 33.4% when the pressure of H_2 is 4Mpa, and the yield of sorbitol and 1,2-PG have different trends, one gradually increases, another comes down little by little. The yiled of other polyols are too low to accurately analyze the trend. According to the yield of EG, the subsequent experiment adopted the H_2 pressure of 4Mpa.

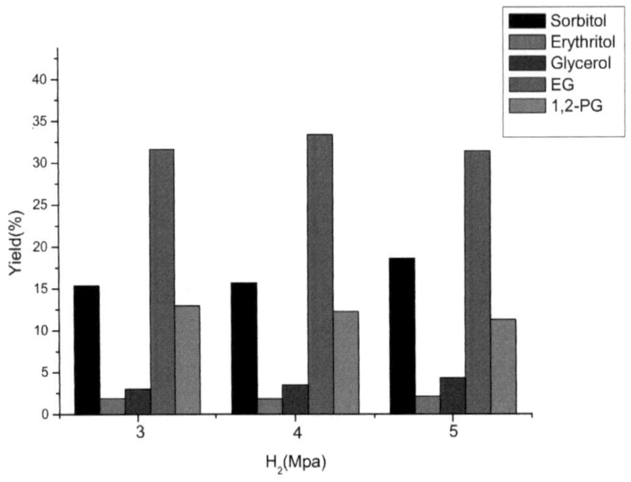

Fig 1 The effect of H_2 pressure on the yield of polyols from cellobiose conversion

1.2 The effect of temperature on the yield of polyols from cellobiose conversion

It was reported that the reaction temperature had great effect both on kind and yield of products[18]. As seen in Fig. 2, with the increase of temperature in the reaction from 220℃ to 240℃, the yield of EG and 1,2-PG have a increasing trend, when the reaction temperature is 240℃, the yield of EG and 1,2-PG are 33.4% and 12.2% respectively. However, the yield of sorbitol are greatly reducing with the increase of temperature. The subsequent experiment adopted the reaction temperature of 240℃.

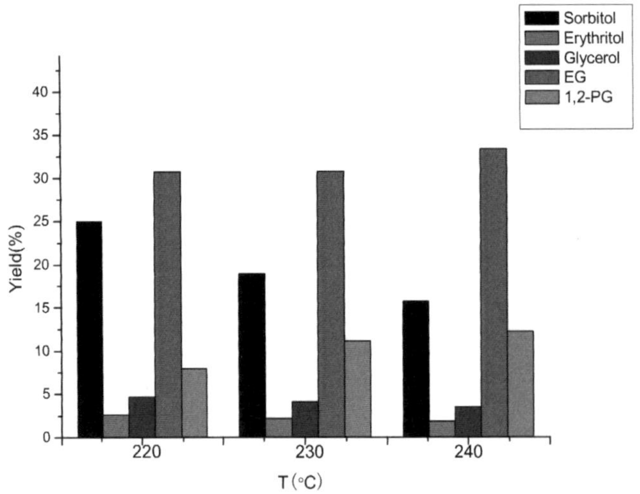

Fig 2 The effect of reaction temperature on the yield of polyols from cellobiose conversion

1.3.The effect of reaction time on the main yield of polyols from cellobiose conversion

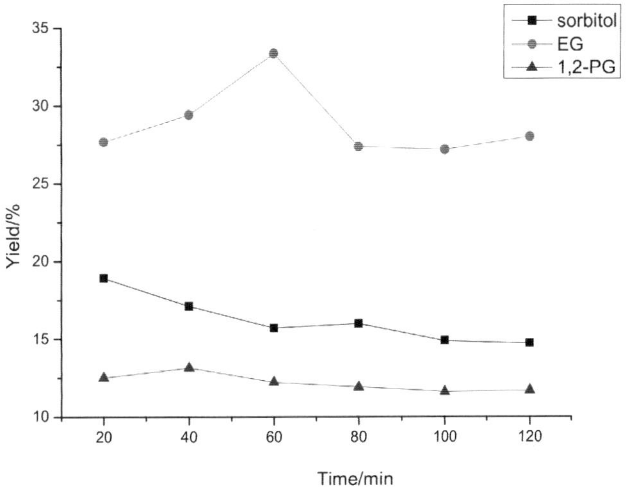

Fig 3 The effect of reaction time on the yield of the main polyols from cellobiose conversion

Six batch of experiments were done with differnet reaction time. As shown in Fig 3, the effect of time on the main yield of polyols from cellobiose conversion are obvious, from 20 min to 120 min, the yield of EG, 1,2-PG, and sorbitol present very different trends. With the increase of reaction time, the highest yield of EG achieved at 60 min, the production of sorbitol is on the decline, while the yield of 1,2-PG is about 12% all the time.

2 The cleavage of bonds and the products in the reaction

During the whole catalytic hydrogenation, the cleavage of different bonds and the production of various chemicals are involved. Broadly speaking, the catalytic hydrogenation of cellobiose involves the breakage of both C-O and C-C bonds. It is reported that the C-O cleavage proceeds through hydrolysis or dehydration or hydrogenation, whereas the C-C breaking occurs via the retro-aldol reaction[19-20], various chemicals are produced through catalytic hydrogenation.

As seen in Table 1, depending on the reaction temperature, the product varies from monose to polyol. With the increase of temperature from 120℃ to 240℃, the conversion of cellobiose reaches up to 99.0% from 27.9%, glucose and fructose undergo the catalytic hydrogenation with almost a 100% conversion at 240℃. When the reaction temperature below 150℃, glucose is producted from hydrolysis of cellobiose over an acid catalyst(phosphotungstic acid) needing a long reaction time, and trace of acetol, fructose, sorbitol, EG, 1,2-PG, erythritol and glycerol are produced, it is reported that the existence of fructose can attribute to the isomerization of glucose especially in presence of activated carbon, which promotes the isomerization of glucose into fructose by its surface basicity[21] and 1,2-PG forms via the complex pathway from fructose as the precursor[22]. Then, with the increase of reaction temperature from 150℃ to 180℃, sorbitol is produced from hydrolytic hydrogenation of cellulose under the presence of pressurized H_2 and hydrogenation catalyst(Ru/C), sorbitol is quite stable and not converted into EG under the present reaction conditions[11]. Besides, when the reaction proceeds at temperature above 210℃, EG and 1,2-PG become the major product in the presence of multifunctional catalysts[14], other polyols such as glycerol, 1,2-butanediol, erythritol, mannitol, etc are produced as by-products. Meanwhile, when cellobiose is replaced by glucose or fructose as feedstock, at the temperature above 210℃, the higher yields of EG and 1,2-PG are formed from the conversion of glucose and fructose, respectively. This part of reaction is very complex. Hence, in order to clearly understand the conversion of cellulose or hemicellulose into polyols, the possible pathways of producing polyols in the reaction are needed.

Table 1 The conversion of cellobiose, glucose and fructose at different temperatures

T/℃	Con %	Glu %	Fru %	Sor %	EG %	1,2-PG %	Acetol %	Ery %	Gly %
120[1]	27.9	trace	0	0	0	0	0	0	0
150[1]	44.4	9.0	trace	trace	trace	trace	trace	trace	trace
180[1]	69.8	11.6	4.3	6.4	9.2	2.0	2.3	trace	4.4
210[1]	98.1	trace	trace	11.0	18.7	3.0	4.0	1.4	3.8
240[1]	99.0	trace	trace	9.8	22.3	11.3	trace	2.2	1.7
210[2]	90.3	9.7	0	22.0	20.5	10.1	5.2	1.6	2.7
240[2]	92.9	7.1	0	12.5	16.5	13.5	trace	1.1	1.9
210[3]	100	0	0	-	2.5	19.1	4.0	0	4.0
240[3]	100	0	0	-	3.0	20.3	trace	0	3.7

Reaction conditions: 0.3g cellobiose, glucose or fructose, the same amount of catalyst(5wt%Ru/C), 30mL extremely low phosphotungstic acid(0.01wt%) in an 100mL autoclave and stirred at 4Mpa H_2, and 600rpm for the same time.

Con, Glu, Fru, Sor, EG, PG, Ery, Gly are abbreviations of conversion, glucose, fructose, sorbitol, ethylene glycol, 1,2-propylene glycol, erythritol and glycerol, respectively.

[1] Feedstock is cellobiose.

[2] Feedstock is glucose.

[3] Feedstock is fructose.

3 Possible pathways of the main products from cellobiose conversion

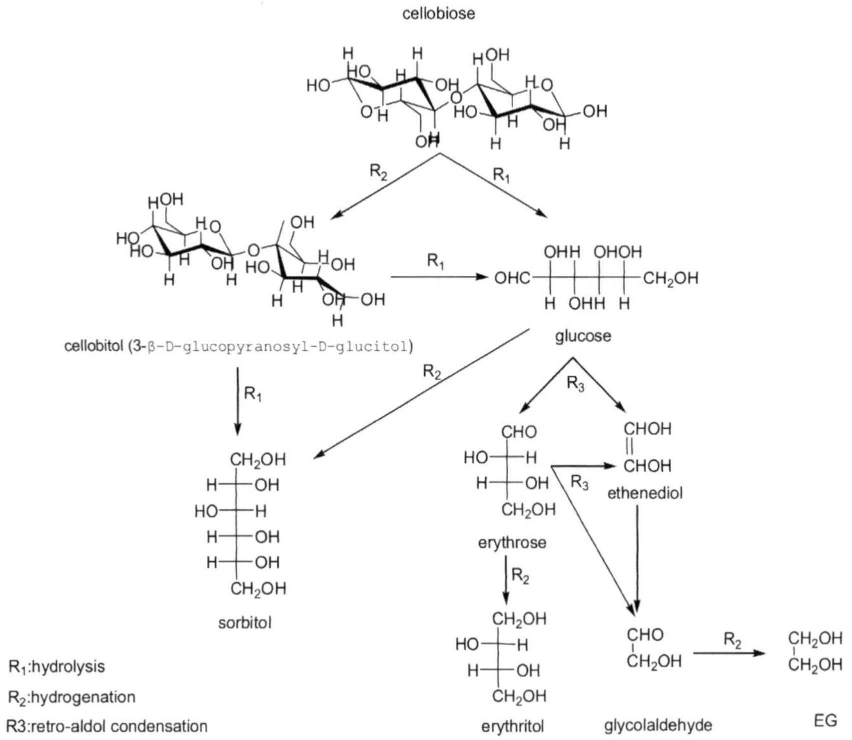

Scheme 1 The possible pathway of producing EG for cellobiose transformation over Ru/C catalyst in PTA solution

According to the results from catalytic process in presence of H_2, cellobiose undergoes hydrolysis to glucose under the catalysis of H^+ which is released from both hot water and dissolved PTA(Scheme 1). Meanwhile, on the one hand, glucose could be directly catalyzed by precious-metal catalyst (Ru/C) to produce sorbitol. On the other hand, with the increase of temperature, the C-C bond cleavage of glucose occurs via retro-aldol reaction under the dissolved tungsten from PTA, erythrose and ethenediol are produced in the reaction. For one thing, erythritol is formed by hydrogenation of erythrose in the presence of pressurized H_2 and hydrogenation catalyst (Ru/C). For another thing, the retro-aldol condensation reaction takes place, ethenediol and glycolaldehyde are produced from the C-C cleavage of erythrose. Together, ethenediol is not stable, but is easily isomerized to the glycolaldehyde which is converted to EG over hydrogenation catalyst in presence of H_2.

Besides, duo to the low concentration of H^+ from extremely low phosphotungstic acid solution, part of glycosidic bonds did not crack, cellobiose was found to undergo hydrogenation of the C-O bond on one of the glucose rings leading to cellobitol(3- β -D-glucopyranosyl-D-glucitol), which was confirmed by several researchers[17,24], and $C_{12}H_{24}O_{11}$ was also detected by UPLC-ESI-MS in our experiment (in Fig 4). In the subsequent reaction, cellobitol would be converted into sorbitol and glucose by hydrolysis. Meanwhile, cellobiose also can undergo retro-aldol reaction on one of the glucose rings to produce 2- β -D-glucopyranosyl-D-erythrose ($C_{10}H_{18}O_9$) and ethenediol. 2- β -D-glucopyranosyl-D-erythrose then was converted into 2- β -D-glucopyranosyl-D-erythritol ($C_{10}H_{20}O_9$), and $C_{10}H_{20}O_9$ was also detected by UPLC-ESI-MS (in Fig 5), the existence of 2- β -D-glucopyranosyl-D-erythritol would further confirm the possible pathway of producing EG in Scheme 1. Whether $C_{12}H_{24}O_{11}$ or $C_{10}H_{20}O_9$ should further confirm chemical structure, this part of experiment is being done.

In a word, the possible pathway of producing EG in the catalytic hydrogenation of cellobiose is proposed according to the results from our experiment and other groups' research progress[22,25]. Under the cooperation between homogeneous tungsten compound (PTA) and heterogeneous precious-metal catalyst (Ru/C), a high yield of EG can be obtained from cellobiose in one pot[12].

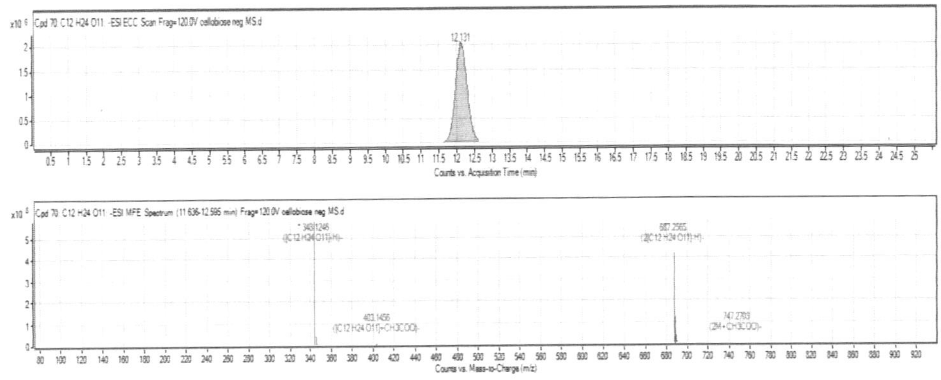

Fig 4 The spectrogram of cellobitol from UPLC-ESI-MS for cellobiose transformation over 5wt% Ru/C catalyst at 180℃, 4Mpa H_2, 60min, and 30mL 0.01wt% phosphotungstic acid solution

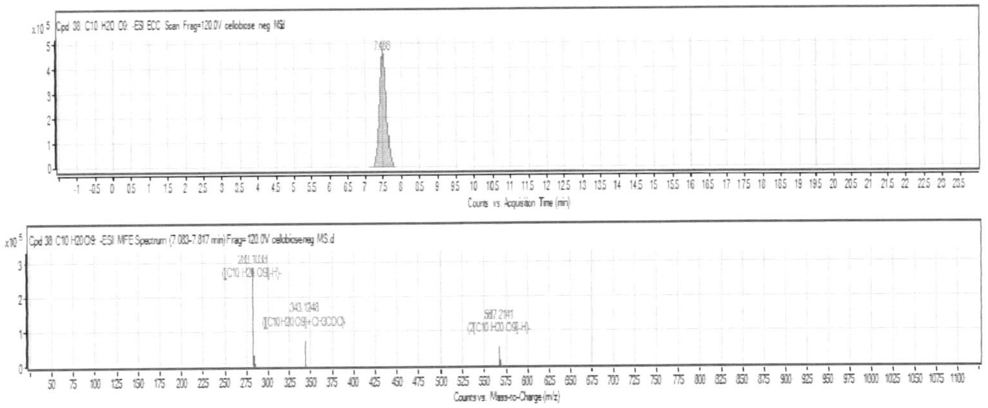

Fig 5 The spectrogram of $C_{10}H_{20}O_9$ from UPLC-ESI-MS for cellobiose transformation over 5wt% Ru/C catalyst at 180℃, 4Mpa H_2, 60min, and 30mL 0.01wt% phosphotungstic acid solution

Besides, glucose would be partly isomerized to fructose especially in presence of activated carbon, the pathway of fructose is different in the reaction. According to a paper from Zhang Tao and coworkers[22], the retro-aldol reaction of fructose produced glyceraldehyde and 1,3-dihydroxyacetone which were detected by HPLC. Then, to further find the main precursor for the formation of 1,2-PG, glyceraldehyde and 1,3-dihydroxyacetone were used as the feedstock in the reaction respectively or together, the result strongly suggested that glyceraldehyde was the main precursor for the formation of glycerol, whereas 1,3-dihydroxyacetone was responsible for the formation of 1,2-PG. While, Liu Hai-chao and coworkers[25] found that glyceraldehyde and propenetriol were produced from the retro-aldol reaction, and propenetriol was not stable, and it was easy to convert into glyceraldehyde or 1,3-dihydroxyacetone in water, but 1,3-dihydroxyacetone was not detected in the reaction. Besides, if 1,2-PG was formed from 1,3-dihydroxyacetone, the hydroxy on 1,3-dihydroxyacetone has the same chemical activity duo to the symmetrical structure, so both C_1 and C_3 on 1,3-dihydroxyacetone can become the C on methyl of 1,2-PG. While, when $^{13}C_1$- fructose and $^{13}C_6$-fructose were used as feedstock, all the C on methyl of 1,2-PG were detected to be ^{13}C after reaction. Therefore, it is glyceraldehyde rather than 1,3-dihydroxyacetone that is the main precursor for the production of 1,2-PG.

According to the conclusions above, there are disputes on the existence of 1,3-dihydroxyacetone and the main precursor for the formation of 1,2-PG in the reaction. Hence, in order to resolve disputes, we studied the possible precursor of 1,2-PG in the reaction.

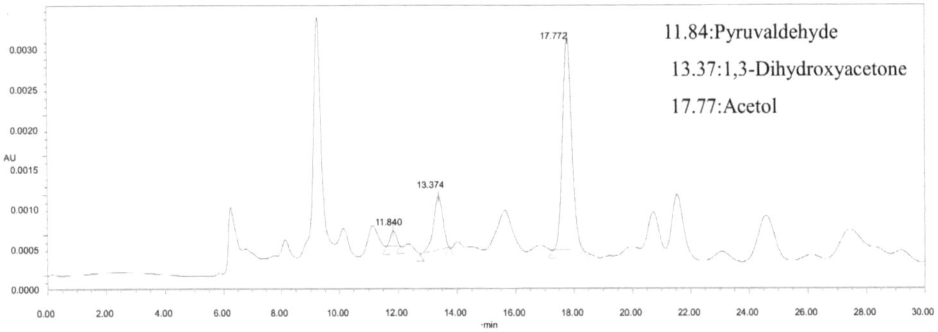

Fig 6 Liquid chromatogram by UV detector(254nm) for cellobiose transformation over 5wt% Ru/C catalyst at 180℃, 4Mpa H_2, 60min, and 30mL 0.01wt% phosphotungstic acid solution

For one part, 1,3-dihydroxyacetone was detected from the products of reaction by HPLC in Fig 6. However, it does not mean that 1,3-dihydroxyacetone is responsible for the formation of 1,2-PG. For another part, 1,3-dihydroxyacetone was used as substitution of cellobiose in the reaction at different temperatures from 120℃ to 240℃, and the results are shown in Fig 7. The highest yield of 1,2-PG is only 2.2% when 1,3-dihydroxyacetone was used for reaction at 240℃. Therefore, it is not easy for 1,3-dihydroxyacetone to produce 1,2-PG in present reaction conditions.

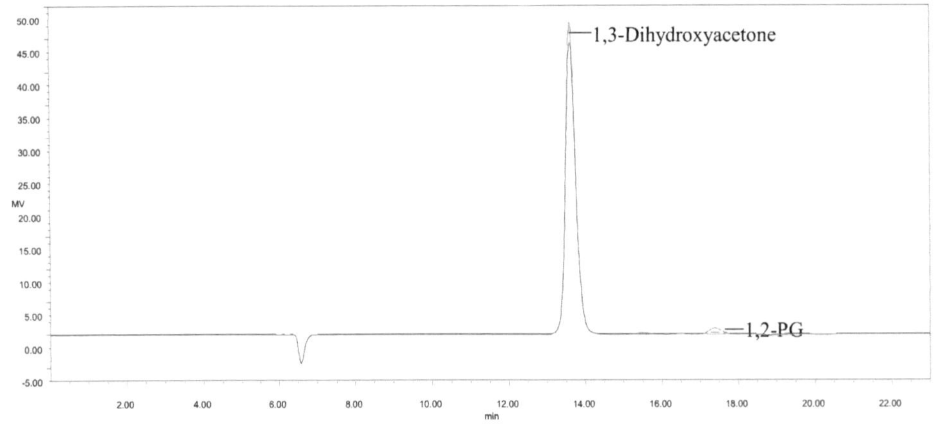

Fig 7 Liquid chromatogram by RI detector for 1,3-dihydroxyacetone transformation over 5wt% Ru/C catalyst at different temperatures from 120 to 240℃, 4Mpa H_2, 60min, and 30mL 0.01wt% phosphotungstic acid solution

Combining the research results from other groups and our own group, we can conclude that the main precursor for the formation of 1,2-PG is not 1,3-dihydroxyacetone in our reaction system. Hence, on the one hand, glyceraldehyde is responsible for the formation of glycerol through catalytic hydrogenation. On the other hand, glyceraldehyde is a precursor for the formation of 1,2-PG via cross–dehydration–hydrogenation reaction pathway (Scheme 2), pyruvaldehyde and acetol are produced as intermediates in Fig 6. However, it does not mean that the conclusions from the groups of Zhang Tao and Liu Hai-chao are not correct, there are differences from conditions to catalysts of the three reactions. Therefore, the main precursor for the formation of 1,2-PG may depend on catalysts and conditions in the reaction, and further researches on the pathways of producing 1,2-PG from cellobiose conversion are being done.

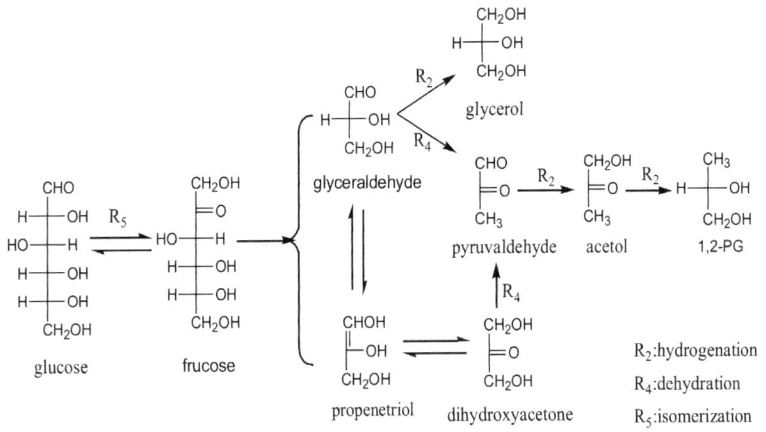

Scheme 2 The possible pathway of producing 1,2-PG for cellobiose transformation over Ru/C catalyst in PTA solution

Conclusion

Using cellobiose as feedstock to converted into polyols in the presence of hydrogen and precious-metal catalyst of Ru/C over extremely low phosphotungstic acid, the yields of ethylene glycol (EG) and 1,2-propylene glycol (1,2-PG) were obtained up to 33.4% and 12.2% respectively at 240 ℃ for 1 h, which had a definite guidance function for catalytic conversion cellulose. As the simplest and basic unit of cellulose, cellobiose is an ideal feedstock for the study of the possible reaction pathways of producing polyols. Combining the research progress from other groups, the detection of intermediates such as pyruvaldehyde, acetol and 1,3-dihydroxyacetone by HPLC and UPLC-MS verified the possible reaction pathways of producing polyols. Besides, glyceraldehyde is a mian precursor for the formation of 1,2-PG via cross–dehydration–hydrogenation reaction pathway rather than 1,3-dihydroxyacetone in the present reaction conditions.

Acknowledgments

This work was supported by the State Natural Sciences Foundation (31270635), the National High Technology Research and Development Program of China (No.2012AA101806) and the Fundamental Research Funds for the Central Universities (2014ZP14)

Reference

[1]Jiying Sun and Haichao Liu.Selective hydrogenolysis of biomass-derived xylitol to ethylene glycol and propylene glycol on supported Ru catalysts[J].Green Chem.,2011,13:135–142.

[2]D.M. Alonso, J. Q. Bond and J. A. Dumesic.Catalytic conversion of biomass to biofuels.Green Chem, 2010,12:1493–1513

[3]D. L. Klass.Biomass for Renewable Energy, Fuels, and Chemicals, Academic Press, San Diego, 1998.
[4]G.W. Huber, S. Iborra, A. Corma.Synthesis of Transportation Fuels from Biomass: Chemistry, Catalysts, and Engineering.Chem. Rev. 2006,106:4044–4098.
[5]Mizuho Yabushita,Hirokazu Kobayashi,Atsushi Fukuoka.Catalytic transformation of cellulose into platform chemicals[J].Applied Catalysis B,2014,145:1-9.
[6]C. Luo, S.Wang, C. H. Liu.Cellulose Conversion into Polyols Catalyzed by Reversibly Formed Acids and Supported Ruthenium Clusters in Hot Water[J].Angew. Chem. 2007, 119:7780–7783.
[7]Fukuoka, A. Dhepe, P. L. Catalytic Conversion of Cellulose into Sugar Alcohols[J]. Angew.Chem., Int. Ed. 2006, 45:5161–5163.
[8]Dhepe, P. L.Fukuoka, A. Cellulose Conversion under Heterogeneous Catalysis[J].ChemSusChem,2008, 1:969–975.
[9]Ji, N.Zhang, T.Zheng, M.Wang, A. Wang, H. Wang, X.Chen, J. G. Direct Catalytic Conversion of Cellulose into Ethylene Glycol Using Nickel-Promoted Tungsten Carbide Catalysts[J]. Angew. Chem., Int. Ed.2008, 47: 8510–8513.
[10]Yue, H.Zhao, Y.Ma, X.; Gong, J. Ethylene Glycol: Properties, Synthesis, and Applications.Chem. Soc. Rev. 2012, 41:4218–4244.
[11]Guanhong Zhao,Mingyuan Zheng,Junying Zhang,Aiqin Wang,and Tao Zhang.Catalytic Conversion of Concentrated Glucose to Ethylene Glycol with Semicontinuous Reaction System[J]. Ind. Eng. Chem. Res. 2013, 52:9566−9572.
[12]Zhijun Tai,Junying Zhang,Aiqin Wang,Mingyuan Zheng,Tao Zhang. Temperature -controlled phase-transfer catalysis for ethylene glycol production from cellulose[J].Chem. Commu, 2012,48:7052–7054.
[13]Zhijun Tai et al.Catalytic Conversion of Cellulose to Ethylene Glycol over a Low-Cost Binary Catalyst of Raney Ni and Tungstic Acid[J].ChemSusChem 2013,6:652–658.
[14]A.Y.Yin,X.Y.Guo,W.L.Dai,K.N.Fan. The synthesis of propylene glycol and ethylene glycol from glycerol using Raney Ni as a versatile catalyst[J].Green Chem.2009,11:1514-1516.
[15]Regina Palkovits,et al.Heteropoly acids as efficient acid catalysts in the one-step conversion of cellulose to sugar alcohols[J].Chem.Commun,2011,47:576-578.
[16]Ken-ichi Shimizu,et al.Effects of Brønsted and Lewis acidities on activity and selectivity of heteropolyacid-based catalysts for hydrolysis of cellobiose and cellulose[J]. Green Chem,2009,11:1627–1632.
[17]Leila Negahdar,Regina Palkovits,et al.Kinetic investigation of the catalytic conversion of cellobiose to sorbitol[J].Applied Catalysis B:Environmental,2014,147:677– 683.
[18]Wang Aiqin, Zhang Tao.One-Pot Conversion of Cellulose to Ethylene Glycol with Multifunctional Tungsten-Based Catalysts[J].Accounts of chemical research, 2013, 46 (7):1377-1386.
[19]B. M. Kabyemela, T. Adschiri, R. Malaluan, K. Arai.Degradation Kinetics of Dihydroxyacetone and Glyceraldehyde in Subcritical and Supercritical Water.Ind. Eng. Chem. Res.1997, 36:2025–2030.
[20]B. M. Kabyemela, T. Adschiri, R. M. Malaluan, K. Arai. Glucose and Fructose Decomposition in Subcritical and Supercritical Water: Detailed Reaction Pathway, Mechanisms, and Kinetics.Ind. Eng. Chem. Res.1999, 38:2888 –2895.
[21]Liu Y,Luo C,Liu H C.Tungsten trioxide promoted selective conversion of cellulose into propylene glycol and ethylene glycol on a ruthenium catalyst[J].Angew Chem Int Ed,2012,51(13):3249-3253.
[22]Likun Zhou,Aiqin Wang,Changzhi Li,Mingyuan Zheng,and Tao Zhang. Selective production of 1,2-propylene glycol from Jerusalem Artichoke Tuber using Ni-W2C/AC catalysts[J]. ChemSusChem,2012,5:932-938.
[23]Jifeng Pang, et al.Catalytic Conversion of Concentrated Miscanthus in Water for Ethylene Glycol Production[J]. AIChE Journal,2014,6(60):2254-2262.
[24]Ning Yan, Chen Zhao, Chen Luo, Paul J.Dyson, Haichao Liu, and Yuan Kou. One-Step Conversion of Cellobiose to C_6-Alcohols Using a Ruthenium Nanocluster Catalyst[J].J. AM. CHEM. SOC.2006,128:8714-8715
[25]Yue Liu, Haichao Liu.Mechhanism for solid acid catalysis and selective cleavage of C-Cbonds of sugars in catalytic conversion of cellulose into polyols[D].Peking University, 2013,6:101-107.

375448 Design of Process Plant for the Production of Formaldehyde and Formic Acid from Lignin

Abstract:

The production of platform chemicals from biomass is the current focus of chemical research. Platform chemicals are the precursors of many household and industrial products. These chemicals can be converted into several high-value products and materials including resins, paints, plastics, adhesives, solvents, and synthetic fibers. Most platform chemicals are currently derived from fossil fuels. One potential source to replace petroleum as a precursor for platform chemicals is plant biomass. Biomass is composed of cellulose, hemicellulose, and lignin. The composition of the component of biomass suggests that it could also be utilized in the production of platform chemicals. Most bio-refinery research has focused on converting cellulose and hemicellulose into ethanol. However, lignin, which constitutes 15-25% of plant biomass, is considered a waste product in bio-refineries. In this project, the feasibility of utilizing lignin for the production of formaldehyde and formic acid is considered. The Douglas hierarchy is utilized to develop process flowsheets for making these products from lignin. Several simulations are conducted in the process simulator ASPENPlus to study the technical as well as economic viability of these processes.

Elucidation of cellulose-lignin interaction during pyrolysis: A Py-GC-MS study

Shiliang Wu, Dekui Shen, Rui Xiao, Southeast University, NanJing, China

Abstract

Interaction between the three main components (cellulose, hemicelluloses, and lignin) of biomass is notably observed under pyrolysis condition. Method (both experimental and analytical) for estimating the extent of interactions is required in order to gain the better understanding of the mechanism of pyrolysis of whole biomass system. In this work, two kinds of mixture (direct mixture and table mixture) for cellulose-lignin and levoglucosan-lignin were pyrolyzed in Pyroprobe 5250 coupled with GC-MS under temperature range 500-700 ℃. While there is a slight difference between calculated and experimental volatile and char yield, the significant interactions on the product distribution were observed: the presence of lignin promoted the formation of light-molecular weight compounds from cellulose (such as acetone, glycolaldehyde, acetone alcohol, and butanedial) but inhibited the formation of anhydrosugars, levoglucosan and oligosaccharides. While most of lignin-derived products showed stability under co-pyrolysis with cellulose, excluding some certain compounds (p-cyresol, 2,3-dihydrobenzofuran, guaiacol and 4-vinylguaiacol). A statistical method by correlation r and p-value was adopted to evaluate the interaction strength on product distribution from two mixtures of cellulose-lignin pyrolysis, finding that interactions of table mixture were stronger than direct mixture. Moreover, interaction strength of direct mixture was decreased with increasing temperature, while table mixture achieved the strongest interaction at 600 ℃. For levoglucosan-lignin direct mixture, co-pyrolysis enhanced the levoglucosan-derived products while inhibited the lignin-derived products, revealing that hydrogen transfer and activated intermediates from cellulose and lignin maybe the dominant interaction mechanism. It is believed that the difference of interaction strength for direct and table mixture is caused by the difference of interaction phase. For table mixture, gas phase, liquid phase and even liquid-soild phase interactions are existed. But for direct mixture, only gas phase and limited liquid phase interaction is thought to be existed.

Introduction

As a CO2 neutral and sustainable energy, biomass has received more and more attention for renewable production of fuels, chemicals and energy. In United States and European Union, the renewable energy produced from biomass has surpassed the total combined contribution from hydropower, wind power, geothermal energy, and solar power [1]. In addition to energy, biomass is the only renewable energy for the production of plat chemicals; besides, it is also the only renewable source for liquid transportation fuel.

Biomass is composed of cellulose, hemicellulose and lignin, with a little amount of extractives and inorganic salts. During pyrolysis, the existence of inorganic salts influences the distribution of volatiles, accelerates the formation of low molecular weight species [2]. A number of studies on pyrolysis of biomass [3,4] and biomass components (cellulose [5,6], hemicelluloses [7] and lignin [8]) were reported in literature. But little attention was paid to the interactions between cellulose, hemicelluloses and lignin, which also played an important role during the pyrolysis of biomass.

Usually, these three components are thought to pyrolyze separately with no interactions between them. Caballero et al. investigated the pyrolysis of almond shell and its components (holocellulose and lignin) by Pyroprobe 1000 within the temperature range from 700 ℃ to 900 ℃, products from pyrolysis were analyzed by gas chromatography, they reported that the yields of gaseous product from almond shell can be obtained from the yields of holocellulose and lignin [9]. Alen et al. used pyrolysis-gas-chromatography with mass-selective detection (Py-GC/MSD) to study the thermochemical behaviour of pine wood and its three components in the temperature range 400-1000 ℃, they remarked that wood appeared to behave as the sum of its constituents [10]. Miller et al. and Rao et al. assumed that there were no interaction between cellulose, hemicellulose, and lignin, so the pyrolysis of general biomass materials could be modelled via a superposition of the kinetics of three components. They stated that experiments showed a favorable agreement with the model predictions [11,12]. Sevenson et al. carried out the pyrolysis of cellulose, birch hemicelluloses, and birch lignin in a single particle reactor at constant reactor temperature in the range 225-650 ℃, the calculated results that the three independent materials added together in the proportions of birch wood [13]. Biagini et al. applied summative law for the TG results to obtain the chemical composition of biomass, where calculated values obtained by this method are in good agreement with the experimental results [14]. Yang et al. performed pyrolysis of synthesized biomass samples and three components by a thermogravimetric analyzer (TGA), suggesting that negligible interaction among the three components was found during the pyrolysis of synthesized samples [15]. Fushimi et al. reported that there was no significant interaction for the gas evolution during pyrolysis of model biomass samples (composed of cellulose, hemicelluloses and lignin) in a continuous cross-flow moving bed type differential reactor [16]. Qu et al. claimed that the additivity law can predict reasonably the trend of product yields of biomass samples from their composition of three components, they conducted fast pyrolysis of samples in a tube furnace between 350-650 ℃ [17]. Yoon et al. validated a three-parallel reaction kinetic model which was formulated on the basis of the weight sum of reaction rates for individual components by using a combined thermogravimetry and gas chromatography [18].

In contrast, some researchers reported that the product yield and kinetics of biomass in pyrolysis cannot be explained by additivity law of the three components because of the interaction among the three components during tar production and evolution. Hosoya et al. pointed out that there were significant interactions in cellulose-lignin pyrolysis at gasification temperature (800 ℃), lignin inhibited the thermal polymerization of levoglucosan formed from cellulose and enhanced the formation of the low molecular weight products, while cellulose reduced the secondary char formation from lignin. Comparatively weak interactions were also observed in cellulose-hemicellulose pyrolysis [19]. Worasuwannarak et al. investigated pyrolysis behaviours of rice straw, rice husk, and corncob by TG-MS technique, indicating obvious interactions between cellulose and lignin during pyrolysis. The interactions lead to a decrease in tar yields but an increase in char yields [3]. Couhert et al. doubted whether additivity law is allowed to predict gas yields of biomass after rapid pyrolysis at high temperature, so experiment in an entrained flow reactor (EFR) was conducted, showing that interactions occurred both outside (i.e. in the gas phase) and inside the mixed samples [20,21]. Wang et al. tested pyrolysis characteristics of biomass by using both a thermogravimetric analyzer coupled with a Fourier-transform infrared spectrometer (TG-FTIR) and an experimental pyrolyzer. By analysing the main products by gas chromatography-mass spectrometry (GC-MS), they considered that pyrolysis characteristics of biomass cannot be predicted through its composition in the main components [22]. Liu et al. investigated the synthetic biomass samples by using the technology of thermogravimetric analyzer coupled with Fourier transform infrared spectrometer (TG-FTIR), revealing the occurrence of interactions among biomass components during pyrolysis [23].

Besides, though some researchers reported there were no interactions between three components by analysising the gas products and yields, they claimed interactions are occurred when other aspects (such as tar products, low temperatures) were analysed. Caballero et al. found that slow thermal decomposition of almond shells cannot be predicted by the sum of lignin and holocellulose [9]. Fushimi et al. observed that the interactions among cellulose, xylan and lignin suppressed the evolution of levoglucosan and significantly increase the evolution rate of 5-methylfurfural [16,24].

Finally, conclusions were different depending on the researchers and the experiment (different reactors, analytical methods and equipments, experiment conditions). Some authors claimed that there was no interaction between the three components when pyrolysised together. But others confirmed that interactions were indeed occurred under pyrolysis when products distribution was analysed. Even for the same author, different conclusions were got when different analytical methods were used. Further studies are necessary for better understanding of biomass pyrolysis. As cellulose and hemicellulose are all polysaccharides, while lignin is composed with methoxylated phenylpropane structures, besides, the interactions between cellulose-hemicellulose are smaller than cellulose-lignin [19], so the interactions between cellulose-lignin are more meaningful than cellulose-hemicellulose

In this article, pyrolyzer together with gas chromatography-mass spectrometer (Py-GC-MS) was used to confirm the interactions between cellulose and lignin in the temperature range 500-700 ℃. The influence of mixture methods for synthetic biomass samples and temperatures were checked. An evaluation method was proposed to assess the cellulose-lignin interaction. Levoglucosan-lignin mixtures were also pyrolysed to shed a light on the mechanism of the cellulose-lignin interaction. Finally, a possible cellulose-lignin interaction mechanism was put forward.

Experiment

Materials

Microcrystalline cellulose power (Sinopharm Chemical Reagent Co.) and levoglucosan (Sigma-Aldrich Co.) were used as commercial samples. Lignin was isolated and purified from black liquor, the detailed method for lignin extraction was shown elsewhere [25].

Cellulose-lignin mixture was prepared by two methods:

(i) Direct mixture: Cellulose and lignin were mixed in equal mass proportion with spoon in a container. Cellulose and lignin powder are all smaller than 100 mesh screen (<150 μm). In this case, interaction in gas phase is favoured, but the interaction between intermediate liquid form cellulose and lignin may be small, because the contact of cellulose-lignin is not very compact, air are existed in the pores of the mixture, which prevent the interaction of cellulose-lignin during pyrolysis.

(ii) Table mixture: Cellulose and lignin were firstly mixed in equal mass as direct mixture, then the miture was tabled by manual hydraulic press machine under 20 Mpa, after that, the tabled mixture was milled to smaller than 100 mesh screen (<150 μm). In this case, as air was removed from the cellulose-lignin mixture, both intermediate liquid phase and gas phase interaction will be expected.

Py-GC-MS

Py-GC-MS was used to investigate the pyrolysis characteristics of cellulose, lignin, cellulose-lignin mixture and levoglucosan-lignin mixture under flash heating mode (heating rate from 10 to 20000 K/s). Fast pyrolysis analyzer (CDS5250) is coupled with the GC-MS instrument (Trace DSQ II) to indicate the distribution of products under the fixed temperature (room temperature~1400 ℃). Only 0.5 mg samples were loaded in the pyrolysis tube before the test to avoid heat diffusion. The experiment was carried out under the temperature of 500 ℃, 600 ℃ and 700 ℃ at the heating rate of 20 ℃/ms with the residence time of 10 s. Due to the poor thermal conductivity of biomass samples and the high heating rate of pyrolyzer, its actual pyrolysis temperature was lower than the set value, reported as 100 ℃ for the difference [26]. The evolved volatiles were analyzed by the couple GC-MS, where the injector temperature was kept at 300 ℃. The chromatographic separation was performed with a TR-35MS capillary column. The oven temperature was programmed from 40 ℃ (3 min) to 200 ℃ (1 min) with 4 ℃/min heating rate, and then to 280 ℃ (1 min) with 20 ℃/min heating rate. The mass spectra were operated in EI mode at 70 eV. The mass spectra were obtained from 35 to 650 (m/z). The chromatographic peaks were identified according to the NIST MS library v2.0. Repeated tests were carried out, finding that the error for absolute peak area was no more than±10%. As calibration has not done for GC-MS, absolute peak area was used to compare the difference of prediction and experiment. Jiang reported that the linearity of peak area and mass for each standard compound was very high (the R^2 values were above 0.98) [27]. So it was receivable for the use of absolute peak area in this article.

Results and discussion

Cellulose-Lignin pyrolysis

Figure 1 shows the formation of volatiles and char of cellulose, lignin, predicted cellulose-lignin mixture without interactions, direct mixture, and table mixture. The yields of char all decreases with increasing temperature. Lignin has the maximum char yield at all conditions except 500 ℃, where char of direct mixture is a little more than lignin. Cellulose is always with the minimum char yield. As for the difference between prediction and experiment, the different is not significant at 600 ℃ and 700 ℃. But at 500 ℃, the char yield of direct mixture is higher than prediction. This may attributes to that air in the direct mixture prevents heat convection, that makes high char yield at low temperatures, when temperature increased, the hinder of air becomes insignificantly, so there is no significant difference between prediction and experiment.

Table 1 summarizes the product distribution of cellulose-lignin pyrolysis. Cellulose, lignin, direct mixed cellulose-lignin, and table mixed cellulose-lignin are pyrolysised under 500 ℃, 600 ℃, and 700 ℃, respectively. To evaluate the difference between predicted yields by additive law and experiment yields of individual products from mixtures, yield ratio γ_i is used:

$$\gamma_i = \frac{Y^i_{mix}}{x_{cel} Y^i_{cel} + x_{lig} Y^i_{lig}} \qquad (1)$$

In (1), Y_{mix} is the experimental mixture yield (for direct mixture cellulose-lignin or table mixture cellulose-lignin, they are marked withγ_D andγ_T, respectively) while Y_{cel} and Y_{lig} are the yields from single component pyrolysis of cellulose and lignin, respectively. x_{cel} and x_{lig} are the initial mass fractions, in this article, $x_{cel} = x_{lig} = 0.5$. Aγ_i value of 1 for individual product i indicates that its production is not affected by co-pyrolysis of cellulose and lignin, while value less or greater than 1 depict product inhibition or promotion, respectively.

For cellulose-derived compounds, the formation of furanone, furfural, levoglucosan, DAGP (1:4;3:6-dianhydroglucopyranose), and 5-hydroxymethylfurfural are all inhibited when lignin is mixed with cellulose, but for light weight compounds, such as acetone, glycolaldehyde, acetone alcohol, butanedial, and methyl pyruvate, are promoted. Besides, when cellulose-lignin co-pyrolysised at 500 ℃, oligosaccharides formed when cellulose pyrolysised alone can not be found. Lignin may promote the secondary reaction of cellulose so that the yields of oligosaccharides, anhydrosugars, levoglucosan and other intermediates decrease while the yields of light weight compounds increase. Piskorz et al. reported that the formation of light weight compounds (such as glycolaldehyde) and anhydrosugars (such as levoglucosan) are through a competition mechanism [28]. And it is observed that some inorganic substances and higher temperatures will reduce the yield of levoglucosan and enhance the formation of glycolaldehyde and hydroxyacetone [2]. Here, for the co-pyrolysis of cellulose and lignin, the role of co-pyrolysis is like some inorganic salts and higher temperatures, accelerating the reactions that led to the formation of low molecular weight species from cellulose as compared to those leading to anhydrosugars. This results is different from Hosoya's experiment [19,29], where both the light weight compounds and anhydrosugars are promoted.

For lignin-derived compounds, no such phenomena as cellulose are found. But phenol,4-methyl-, bezofuran,2,3-dihydro-, 2-methoxy-4-vinylphenol, and 3-tert-butyl-4-hydroxyanisole are all promoted when co-pyrolysised. Besides, phenol,

2-methoxy-4-methyl- (guaiacol) is enhanced significantly under 600 ℃. While other compounds don not show a consistent tendency for different mixture methods and temperatures, but the yield ratios for them are near 1, showing invariableness under co-pyrolysis, this may attribute to the stability of benzene ring. Hosoya et al. reported that cellulose enhanced the formation of some lignin-derived products including guaiacol, 4-methyl-guaiacol and 4-vinyl-guaiacol, while effects on other products were small[19],which is just consistent with our experiment results.

Strength of interactions evulation

Though there are some experiments have been carried on interactions of the three components, there are no efficient methods to analysis the results. It is usually by comparing the results (gas yields, product distribution of tar, TG-DTG curves) of prediction and experiment of co-pyrolysis to conclude whether interaction is occurred during co-pyrolysis or not, this method is arbitrary. That is why results are not in agreement for different researchers.

In this article, statistical method is used to evaluate the strength of interactions. For random variables X and Y, the expected values for them are E(X) and E(Y), respectively. So, covariance of X and Y could be presented as follows:

$$Cov(X,Y)=E[(X-E(X))(Y-E(Y))] \quad (2)$$

Where, Cov(X,Y) is the covariance of X and Y. If Cov(X,Y)=0, it means that random X and Y is mutual independence.

The variances of X and Y are denoted as D(X) and D(Y). Then, correlation coefficient r is defined based on variances and covariances:

$$r = Cov(X,Y)/\sqrt{D(X)D(Y)} \quad (3)$$

The value of r is between -1 to 1, if X and Y is positive correction, the value for r is 1. The bigger the absolute value of r, the correlation of X and Y is more intimate. r is used to evaluate the interactions between cellulose-lignin.

Besides, p-value is also used to check whether the difference between prediction and experiment is caused by statistical discrepancy or interactions.

Figure 2 presents the comparison of pyrolysis products between prediction and experiment. The x-axis shows the prediction values for the absolute peak area of products after co-pyrolysis. The values for the absolute peak area are took the logarithm to the base 10 for briefness. The absolute peak area is normalized with mass (so the unit for product yield is mg^{-1}). This treatment is also conducted with y-axis, which shows the experiment value. So if there is no interaction between cellulose-lignin, the data points should be located on the blue line y=x. To objectively analysis the results, r and p between prediction and experiment are calculated and presented on

Figure 2. Here, $p = 0$ means that the deviations between prediction and experiment are not caused by statistical discrepancy, maybe it is caused by interactions. And the smaller of correlation coefficient r, the stronger of the interactions.

At 500 ℃, the interactions of direct mixture are weaker than table mixture ($r_D = 0.773$ to $r_T = 0.669$), when temperature increased to 600 ℃, this difference amplifies, as $r_D = 0.841$ for direct mixture and $r_T = 0.635$ for table mixture. But at 700 ℃, this difference is very small. This may be explained by the phase of interactions happen. Under low temperature, the interactions occur at liquid phase, as table mixture is more intimate than direct mixture and there is no air in table mixture, so the interactions are stronger for table mixture. While under high temperature, the interactions are mainly happened at gas phase, so the difference is small for direct mixture and simple mixture.

When temperature increased form 500 ℃ to 700 ℃, the strength of interactions for direct mixture is becoming weak. But for table mixture, the strongest is achieved at 600 ℃, not 700 ℃. This attribute to that the interactions are mainly happened at liquid phase and gas phase, and interactions at liquid phase are stronger than gas phase, as products in gas phase are light weight compounds that reactions between them are very small. But for products in liquid phase, there are many intermediates derived from cellulose and lignin, the reactions between them (such as hydrogen transfer) are very vigorous. At 500 ℃, the interaction are mainly happened at liquid phase, but at 500 ℃, a large amount of char is also formed (Figure 1), which limits the interaction at liquid phase. But under 600 ℃, it is reported that the yield of bio-oil is largest [26], so the interactions are focused on liquid phase. When temperature increased to 700 ℃, though there are interactions at liquid phase, but the main products are changed for bio-oil to gas, so the interactions between cellulose-lignin are weakened. The reason for the trend difference of direct and table mixture with temperature is that there is air in direct mixture and the contact of direct mixture is loose, so at 600 ℃, the interactions for direct mixture may be gas interactions not liquid interactions.

So we can conclude that temperature and mixture methods all influence the interactions between cellulose and lignin. The increasing temperature weakens these interactions, and the more intimate the mixture method, the stronger the interactions.

Levoglucosan-lignin pyrolysis

During the cellulose-lignin pyrolysis, levoglucosan is greatly reduced. Levoglucosan is regarded as the intermediate of cellulose pyrolysis, which could polymerize to polysccaride or crack into low molecular products and gas [30-32]. To shed a light on the interaction mechanism of cellulose-lignin, levoglucosan and lignin are mixed by table mixture method (mass ration is also 1:1) and pyrolysised at 600 ℃, where the strongest interactions are found for table mixture.

Table 2 summarizes the products for levoglucosan-lignin pyrolysis. It is interesting to note that levoglucosan-derived products are all promoted, while lignin-derived products are all hindered. The significant promotion of the levoglucosan-derived products are mainly 3-buten-2-ol,

Acetol, 3-cyclopentene-1-acetaldehyde, 2-oxo-, and 2,5-dimethyl-4-hydroxy-3(2H)-furannone. These products could be regarded as decomposition products of levoglucosan. While anhydrosugars and polysccarides (3,4-anhydro-d-galactosan, 2,3-anhydro-d-mannosan and melezitose) are not promote significantly. Which means that lignin promotes the decomposition of levoglucosan. In the research for the secondary reaction of levoglucosan at 400 ℃, Kawamoto et al. concluded that the secondary decomposition of levoglucosan was induced by the lignin-derived products [30]. Which just verifies the results of levoglucosan-lignin pyrolysis at 600 ℃. The char for mixture (5.45%) is largely reduced compared with lignin (29.41% for lignin, though there are char formed on the wall of the quartz tube, the char are so little to be quantized by analytical balance). So the decrease of lignin-derived products may attribute to decomposition of primary products to gas, while gas can't be checked by GC-MS.

After co-pyrolysis of levoglucosan-lignin, the tar from lignin is decreased while gas is increased. This phenomenon is like the effect of increasing temperature, so the roles of interaction and increasing temperature on the lignin-derived products from levoglucosan-lignin pyrolysis are checked in Figure 3. The x-axis is the retention time of the products got from gas chromatography, the yield ration for increased temperature is product yield at 700 ℃ divided by product yield at 600 ℃. In Figure 3, the product yield ration of interaction is consistent with increased temperature. For compounds that are thermal stability, they are also stable under co-pyrolysis. This suggests that there may be a relationship between interaction and temperature. To further understand this relationship, comparison between interaction and increasing temperature is also conducted to products distribution on cellulse-lignin pyrolysis under 500 ℃ and 600 ℃. The results was show on Figure 4. Though the values of yield ration are different, the variation trends along retention time for interaction and increased time are same, except for cellulose-derived products at 600 ℃ (there is only five products can be used for comparison). This phenomenon indicates that the interaction mechanism has a relation with temperature.

Interaction mechanism of cellulose-lignin pyrolysis

There is limited research on interaction mechanism of cellulose-lignin pyrolysis. Mettler et al. investigated co-pyrolysis of levoglucosan with ^{13}C-labeled fructose or deuterated glucose, indicating that carbon atoms of the levoglucosan-derived products are mainly from levoglucosan, but there is a large portion of hydrogen atoms originated from deuterated glucose [33]. Hydrogen transfer plays an important role in co-pyrolysis. Hosoya et al. studied thermal stabilization of levoglucosan in aromatic substances, they concluded the hypothesis of the stabilizing mechanism through CH-π interaction, suggesting that the stability of levoglucosan-aromatic compound complexes is based on the π-electron densities [34]. So under temperature range 500 ℃ to 700 ℃, radicals are possible existed, hydrogen transfer between radicals and sugar-aromatic complexes formed with these radicals will all exist in intermediate pool from cellulose and lignin. A proposed interaction mechanism is presented in Figure 5. After co-pyrolysis, an intermediate pool at liquid phase is formed, where many activated complexes are formed between cellulose-derived and lignin-derived compounds, hydrogen transfer between them is also vigorous, the cellulose-derived products and lignin-derived products are formed form intermediate pool. By this process, products distribution from co-pyrolysis is different from prediction. Though gas phase intraction is also presented, but the residence time of pyrolyser is only 10s, the effect of gas phase interaction is weak than liquid phase interaction.

It is noted that there is a link between interaction and increased temperature mechanism. An increase in temperature corresponds to faster motion of molecules and more intermolecular collisions. For interaction, the reactions of radicals from co-pyrolysis are vigorous than pyrolysised alone due to the existence of hydrogen transfer and activated complexes. The effect of this is like increasing temperature makes radical reactions livelily. So the difference between direct mixture and table mixture is shown in Figure 6. For direct mixture, only gas phase interaction and limited liquid phase interaction are existed, so the intermediate pool for direct mixture is limited, that makes the whole interaction of co-pyrolysis weak. While for table mixture, liquid phase intermediates and tar surround unreacted lignin and cellulose particle, so liquid-soild phase interaction is also existed. This makes char from table mixture is much less than direct mixture at 500 ℃.

Conclusion

Cellulose-lignin pyrolysis is conducted by Py-GC-MS under temperature range 500 ℃ to 700 ℃ with two kind of mixtures (direct mixture and table mixture), indicating that lignin promotes the formation of light weight compounds and inhibits the formation of anhydrosugars, levoglucosan and oligosaccharides. While most of lignin-derived products show stability under co-pyrolysis, with an promotion of phenol,4-methyl-, bezofuran,2,3-dihydro-, 2-methoxy-4-vinylphenol, guaiacol, and 3-tert-butyl-4-hydroxyanisole.

Statistical method is introduced to evaluate the strength of interactions. Interactions between table mixture are all stronger than direct mixture. For direct mixture, the strength of direct mixture is receding with increasing temperature. For table mixture, an maximum interaction is observed at 600 ℃.

Levoglucosan-lignin pyrolysis is also investigated at 600 ℃ for table mixture. Co-pyrolysis enhances the levoglucosan-derived products while prevents the lignin-derived products.

Hydrogen transfer and activated complex in intermediate pool form cellulose-lignin pyrolysis at liquid phase is proposed to be the dominant interaction mechanism.

Reference:

1. Zakzeski J, Bruijnincx PCA, Jongerius AL, Weckhuysen BM(2010), "The catalytic valorization of lignin for the production of renewable chemicals," *Catalysts & Catalysed Reactions.*110(6):3552-3559.
2. Patwardhan PR, Satrio JA, Brown RC, Shanks BH(2010), "Influence of inorganic salts on the primary pyrolysis products of cellulose," *Bioresource Technology.*101(12):4646-4655.
3. Worasuwannarak N, Sonobe T, Tanthapanichakoon W(2007), "Pyrolysis behaviors of rice straw, rice husk, and corncob by TG-MS technique," *Journal of Analytical and Applied Pyrolysis.*78(2):265-271.
4. Shafizadeh F(1982), "Introduction to pyrolysis of biomass," *Journal of Analytical and Applied Pyrolysis.*3(4):283-305.
5. Shen D, Gu S(2009), "The mechanism for thermal decomposition of cellulose and its main products," *Bioresource Technology.*100(24):6496-6504.
6. Shen D, Xiao R, Gu S, Luo K(2011), "The pyrolytic behavior of cellulose in lignocellulosic biomass: a

review," *RSC Adv.*;1(9):1641-1660.

7. Shen D, Gu S, Bridgwater AV(2010), "Study on the pyrolytic behaviour of xylan-based hemicellulose using TG-FTIR and Py-GC-FTIR," *Journal of Analytical and Applied Pyrolysis.*87(2):199-206.
8. Shen D, Gu S, Luo K, Wang S, Fang M(2010), "The pyrolytic degradation of wood-derived lignin from pulping process," *Bioresource Technology.*101(15):6136-6146.
9. Caballero J, Font R, Marcilla A(1996), "Comparative study of the pyrolysis of almond shells and their fractions, holocellulose and lignin. Product yields and kinetics," *Thermochimica Acta.*276:57-77.
10. Alén R, Kuoppala E, Oesch P(1996), "Formation of the main degradation compound groups from wood and its components during pyrolysis," *Journal of Analytical and Applied Pyrolysis.*36(2):137-148.
11. Miller R, Bellan J(1997), "A Generalized Biomass Pyrolysis Model Based on Superimposed Cellulose, Hemicelluloseand Liqnin Kinetics," *Combustion science and technology.*126(1-6):97-137.
12. Rao TR, Sharma A(1998), "Pyrolysis rates of biomass materials," *Energy.*23(11):973-978.
13. Svenson J, Pettersson JBC, Davidsson KO(2004), "Fast pyrolysis of the main components of birch wood," *Combustion science and technology.*176(5-6):977-990.
14. Biagini E, Barontini F, Tognotti L(2006), "Devolatilization of biomass fuels and biomass components studied by TG/FTIR technique," *Industrial & engineering chemistry research.*45(13):4486-4493.
15. Yang H, Yan R, Chen H, Zheng C, Lee DH, Liang DT(2006), "In-depth investigation of biomass pyrolysis based on three major components: hemicellulose, cellulose and lignin," *Energy & fuels.*20(1):388-393.
16. Fushimi C, Katayama S, Tsutsumi A(2009), "Elucidation of interaction among cellulose, lignin and xylan during tar and gas evolution in steam gasification," *Journal of Analytical and Applied Pyrolysis.*86(1):82-89.
17. Qu T, Guo W, Shen L, Xiao J, Zhao K(2011), "Experimental Study of Biomass Pyrolysis Based on Three Major Components: Hemicellulose, Cellulose, and Lignin," *Industrial & Engineering Chemistry Research.* 2011/09/21;50(18):10424-10433.
18. Yoon HC, Pozivil P, Steinfeld A(2011), "Thermogravimetric Pyrolysis and Gasification of Lignocellulosic Biomass and Kinetic Summative Law for Parallel Reactions with Cellulose, Xylan, and Lignin," *Energy & Fuels.* 2012/01/19;26(1):357-364.
19. Hosoya T, Kawamoto H, Saka S(2007), "Cellulose–hemicellulose and cellulose–lignin interactions in wood pyrolysis at gasification temperature," *Journal of Analytical and Applied Pyrolysis.*80(1):118-125.
20. Couhert C, Commandré JM, Salvador S(2009), "Failure of the component additivity rule to predict gas yields of biomass in flash pyrolysis at 950° C," *Biomass and Bioenergy.*33(2):316-326.
21. Couhert C, Commandre JM, Salvador S(2009), "Is it possible to predict gas yields of any biomass after rapid pyrolysis at high temperature from its composition in cellulose, hemicellulose and lignin?," *Fuel.*88(3):408-417.
22. Wang S, Guo X, Wang K, Luo Z(2011), "Influence of the interaction of components on the pyrolysis behavior of biomass," *Journal of Analytical and Applied Pyrolysis.*91(1):183-189.
23. Liu Q, Zhong Z, Wang S, Luo Z(2011), "Interactions of biomass components during pyrolysis: A TG-FTIR study," *Journal of Analytical and Applied Pyrolysis.*90(2):213-218.
24. Fushimi C, Katayama S, Tasaka K, Suzuki M, Tsutsumi A(2009), "Elucidation of the interaction among cellulose, xylan, and lignin in steam gasification of woody biomass," *Aiche J.*55(2):529-537.
25. Hu J, Xiao R, Shen D, Zhang H("Structural analysis of lignin residue from black liquor and its thermal performance in thermogravimetric- Fourier transform infrared spectroscopy," *Bioresource Technology.* (0).
26. Lu Q, Yang X, Dong C, Zhang Z, Zhang X, Zhu X(2011), "Influence of pyrolysis temperature and time on the cellulose fast pyrolysis products: Analytical Py-GC/MS study," *Journal of Analytical and Applied Pyrolysis.*92(2):430-438.
27. Jiang G, Nowakowski DJ, Bridgwater AV(2010), "Effect of the Temperature on the Composition of Lignin Pyrolysis Products," *Energy & Fuels.* 2010/08/19;24(8):4470-4475.
28. Piskorz J, Radlein D, Scott DS(1986), "On the mechanism of the rapid pyrolysis of cellulose," *Journal of Analytical and Applied Pyrolysis.*9(2):121-137.
29. Hosoya T, Kawamoto H, Saka S(2009), "Solid/liquid-and vapor-phase interactions between cellulose-and lignin-derived pyrolysis products," *Journal of Analytical and Applied Pyrolysis.*85(1-2):237-246.
30. Kawamoto H, Morisaki H, Saka S(2009), "Secondary decomposition of levoglucosan in pyrolytic production from cellulosic biomass," *Journal of Analytical and Applied Pyrolysis.*85(1-2):247-251.
31. Hosoya T, Kawamoto H, Saka S(2008), "Different pyrolytic pathways of levoglucosan in vapor-and liquid/solid-phases," *Journal of Analytical and Applied Pyrolysis.*83(1):64-70.

32. Kawamoto H, Murayama M, Saka S(2003), "Pyrolysis behavior of levoglucosan as an intermediate in cellulose pyrolysis: polymerization into polysaccharide as a key reaction to carbonized product formation," *J Wood Sci.*49(5):469-473.

33. Mettler MS, Paulsen AD, Vlachos DG, Dauenhauer PJ(2012), "Pyrolytic conversion of cellulose to fuels: levoglucosan deoxygenation via elimination and cyclization within molten biomass," *Energy & Environmental Science.*

34. Hosoya T, Kawamoto H, Saka S(2006), "Thermal stabilization of levoglucosan in aromatic substances," *Carbohydrate research.*341(13):2293-2297.

FIGURES:

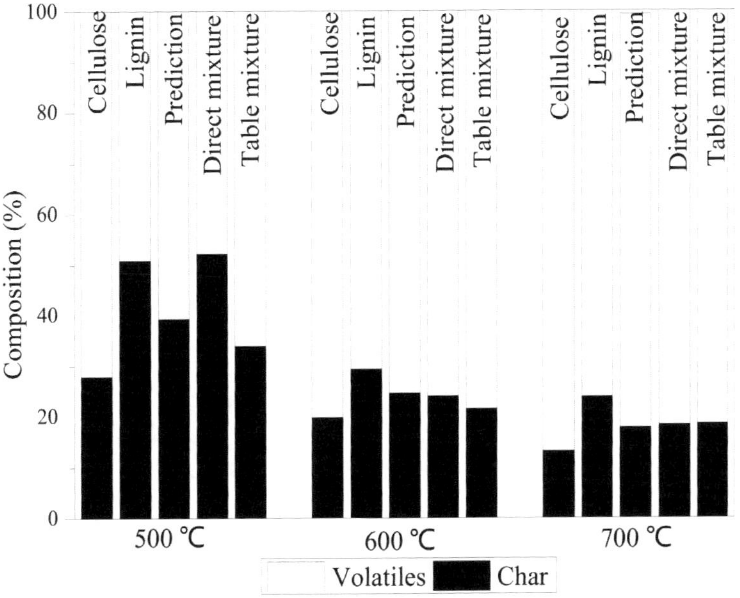

Figure 1. Volatiles and char formation of cellulose-lignin pyrolysis

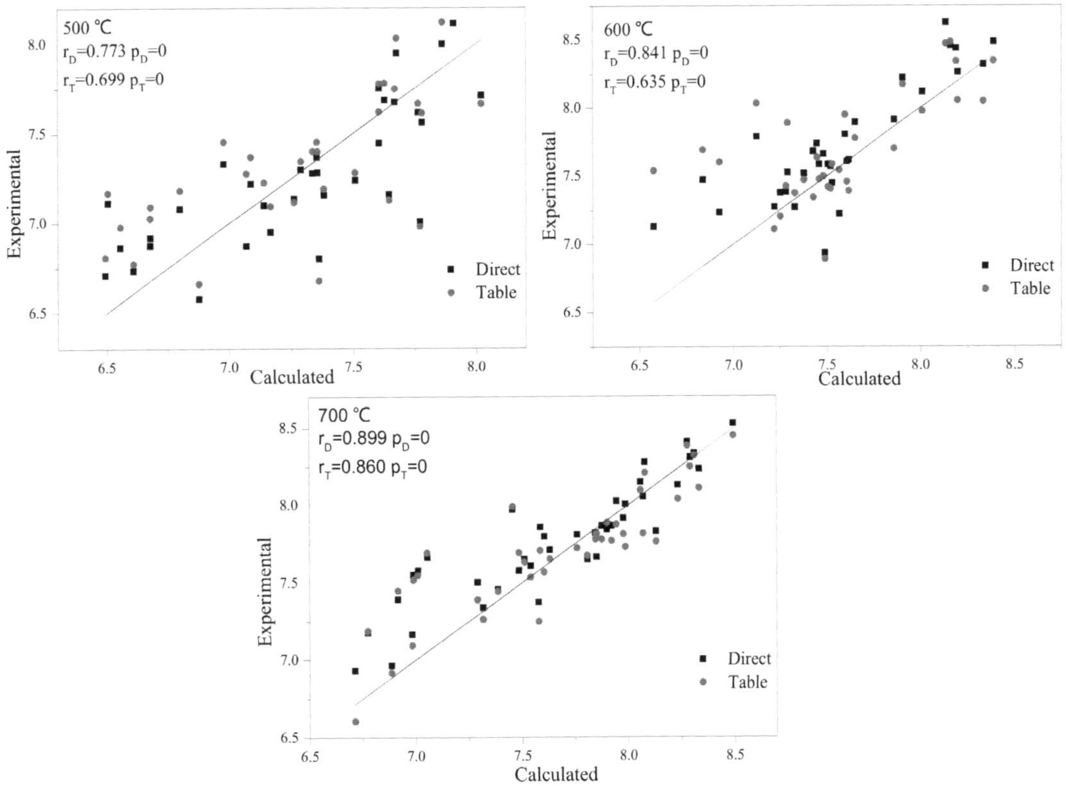

Figure 2. Comparison of pyrolysis products between calculation and experiment for different temperature (500 ℃, 600 ℃, 700 ℃) and mixture methods (direct and table).

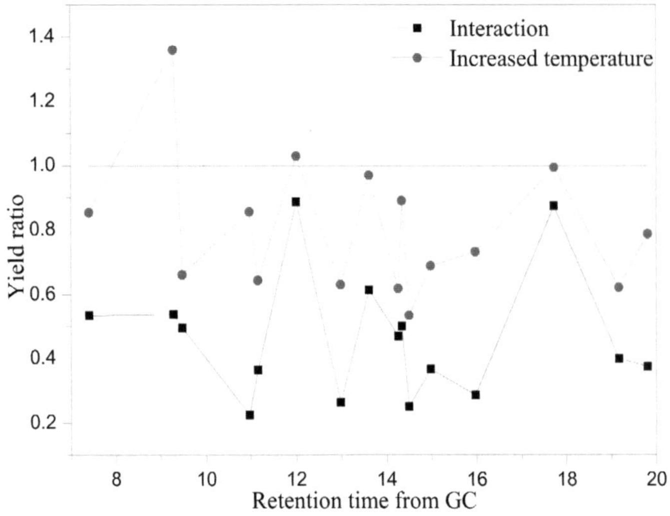

Figure 3. Effect of interaction and increased temperature[a] of 700 on lignin-derived products from levoglucosan-lignin pyrolysis

[a] the yield ration for increased temperature is product yield at 700 ℃ divided by product yield at 600 ℃

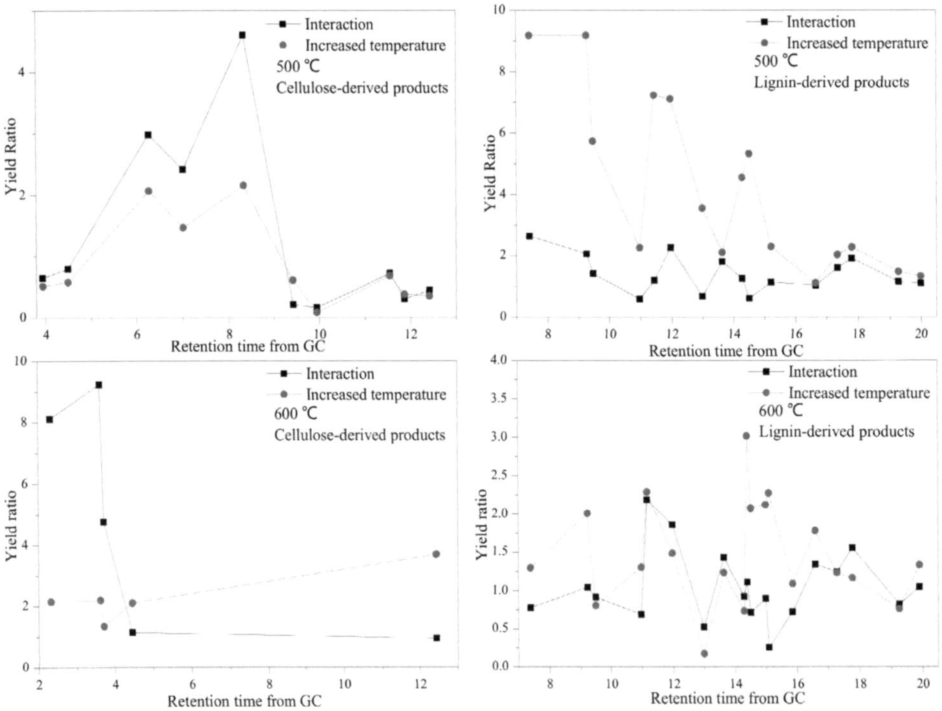

Figure 4. Effect comparison of interaction and increases time on cellulose-lignin products distribuction

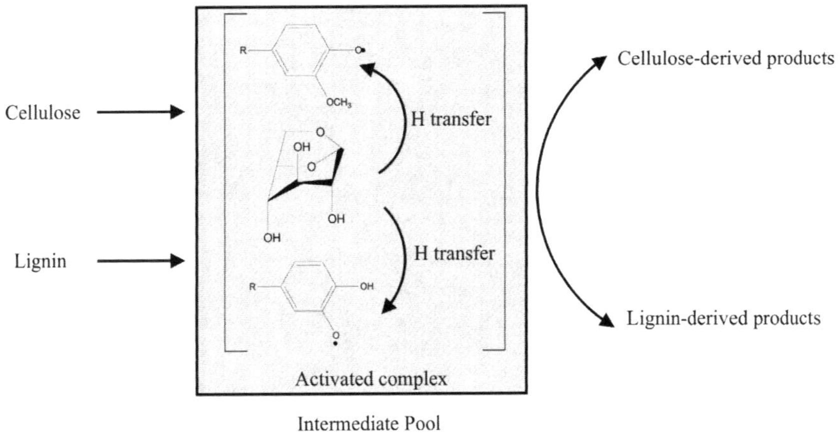

Figure 5. interaction mechanism in intermediate pool of cellulose-lignin pyrolysis

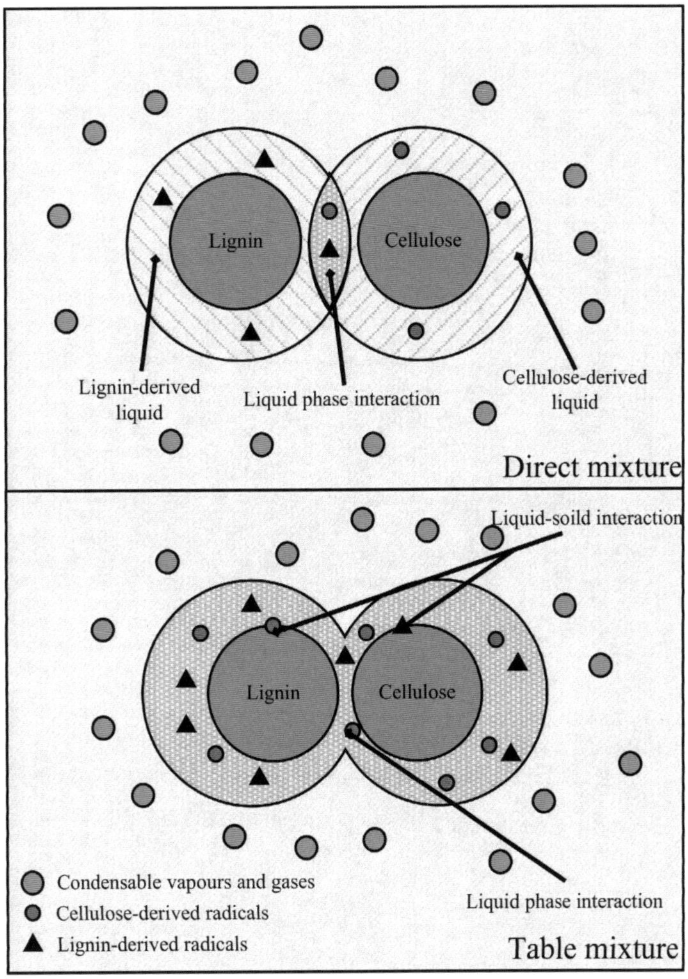

Figure 6. Different interaction phase for direct and table mixture

Table 1. Product distribution of cellulose-lignin pyrolysis under temperature range 500-700 ℃

RT (min)	Compound	500 ℃				600 ℃				700 ℃			
		cellulose[a]	lignin[a]	γ_D	γ_T	cellulose	lignin	γ_D	γ_T	cellulose	lignin	γ_D	γ_T
	Only From Cellulose												
1.611	Acetone	/	/	/	/	/	/	/	/	8.73%	/	1.23	1.09
1.805	Glycolaldehyde	/	/	/	/	1.54%	/	0.99	/	2.34%	/	1.24	1.61
1.866	3-Buten-2-one	/	/	/	/	/	/	/	/	4.90%	/	0.70	0.73
2.304	Acetone alcohol	/	/	/	/	2.19%	/	4.64	8.11	2.18%	/	3.31	3.44
3.613	Butanedial	/	/	/	/	0.62%	/	3.59	9.24	0.63%	/	3.00	3.40
3.713	Methyl pyruvate	/	/	/	/	1.39%	/	2.04	4.76	0.87%	/	4.06	4.34
3.934	Furanone	3.33%	/	0.60	0.65	2.01%	/	/	/	/	/	/	/
4.452	Furfural	8.08%	/	0.71	0.79	5.54%	/	0.84	1.15	5.41%	/	0.65	0.92
6.267	2-Cyclopenten-1-one,2-hydroxy-	1.32%	/	2.25	2.98	3.23%	/	1.72	3.95	2.94%	/	1.87	1.32
6.962	Furfural,5-methyl-	0.87%	/	1.91	2.41	1.52%	/	/	/	1.59%	/	1.06	0.89
8.314	Methylcyclopentenolone	0.44%	/	4.07	4.60	1.13%	/	4.33	7.16	0.79%	/	3.69	3.45
9.939	Levoglucosan	8.21%	/	0.17	0.16	0.80%	/	/	/	/	/	/	/
11.533	DHMP[b]	1.71%	/	0.75	0.72	2.03%	/	0.00	1.82	/	/	/	/
11.866	DAGP[b]	6.14%	/	0.33	0.30	2.73%	/	/	/	3.34%	/	/	/
12.416	5-Hydroxymethylfurfural	14.56%	/	0.49	0.44	9.03%	/	0.46	0.95	7.95%	/	0.50	0.43
	Only From Lignin												
7.375	Phenol	/	0.42%	2.03	2.64	/	1.47%	1.13	0.77	/	1.39%	1.20	1.05
9.262	Phenol,4-methyl-	/	0.37%	1.64	2.06	/	1.27%	1.35	1.04	/	1.86%	1.12	0.92
9.468	Phenol,2-methoxy-	/	4.98%	1.14	1.41	/	10.81%	1.26	0.91	/	6.34%	1.04	0.91
10.613	Phenol,3,5-dimethyl	/	/	/	/	/	0.45%	/	/	/	1.12%	1.17	0.99
10.972	Phenol,4-ethyl	/	3.77%	0.54	0.59	/	3.23%	1.12	0.69	/	3.07%	0.87	0.68
11.161	Phenol,2-methoxy-4-methyl-	/	5.47%	1.01	1.20	/	15%	3.12	2.18	/	10.08%	1.08	0.90
11.831	1,2-Benzenediol	/	/	/	/	/	1.45%	1.07	0.00	/	3.14%	1.05	0.55
11.958	Benzofuran,2,3-dihydro	/	5.57%	1.87	2.26	/	3.58%	2.07	1.86	/	3.88%	1.58	1.35
12.803	1,2-Benzenediol,3-methoxy	/	/	/	/	/	2.14%	1.80	0.83	/	2.85%	1.21	0.85
12.992	Phenol,4-ethyl-2-methoxy-	/	7.09%	0.60	0.68	/	9.45%	0.97	0.52	/	6.96%	0.79	0.60
13.46	1,2-Benzenediol,4-methyl	/	/	/	/	/	/	/	/	/	3.81%	0.96	0.56

13.621	2-Methoxy-4-vinylphenol	/	8.57%	1.36	1.80	/	6.83%	1.78	1.43	/	6.15%	1.34	1.27
14.273	Phenol,2,6-dimethoxy	/	2.64%	1.04	1.26	/	4.54%	1.29	0.92	/	2.43%	0.98	0.80
14.51	1-(4-Methoxyphenyl)-1-propanol	/	0.88%	0.50	0.61	/	1.78%	1.01	0.71	/	2.70%	0.88	0.70
15.209	trans-m-Propenyl guaiacol	/	2.27%	1.02	1.14	/	1.97%	0.28	0.25	/	2.27%	0.94	0.86
15.886	cis-Isoeugenol	/	4.72%	1.40	1.48	/	6.99%	1.16	0.72	/	5.56%	0.79	0.64
16.641	4-Acetyl-2-methoxyphenol	/	4.72%	0.69	1.03	/	1.98%	1.76	1.34	/	2.57%	0.88	0.97
17.332	Guaiacylacetone	/	1.38%	0.63	1.61	/	1.06%	1.39	1.24	/	/	/	/
17.79	3-Tert-butyl-4-hydroxyanisole	/	1.43%	1.36	1.91	/	1.24%	1.96	1.55	/	1.05%	1.38	1.32
19.266	Ethyl citrate	/	2.54%	0.88	1.16	/	1.42%	1.21	0.82	/	0.79%	1.18	1.15
19.983	Methoxyeugenol	/	2.66%	0.84	1.10	/	1.34%	1.52	1.05	/	1.30%	1.56	0.92
						Else							
1.45	CO_2	5.69%	4.72%	1.60	2.08	8.39%	4.14%	1.99	2.12	7.52%	3.41%	1.05	1.02

[a] The column shows the peak area percentage of the compound in volatiles from cellulose or lignin, respectively.

[b] Abbreviations: DHMP – 3,5-dihydroxy-2-methyl-4H-pyran-4-one; DAGP – 1:4;3:6-dianhydroglucopyranose.

Table 2. Product distribution of levoglucosan-lignin pyrolysis at 600 ℃

	Compound	LG	Lignin	LG-L[a]
	Only from LG			
1.625	4-Penten-1-yl acetate	7.21%	/	1.57
1.864	2,3-Butanedione	1.27%	/	2.09
1.936	3-Buten-2-ol	1.36%	/	3.93
2.301	Acetol	2.64%	/	3.01
4.493	3-cyclopentene-1-acetaldehyde,2-oxo-	0.61%	/	2.70
6.227	2-cyclopenten-1-one,2-hydroxy-	1.95%	/	1.79
8.287	2-cyclopenten-1-one,2-hydroxy-3-methyl-	1.87%	/	2.41
9.407	2,5-dimethyl-4-hydroxy-3(2H)-furanone	0.43%	/	5.35
10.083	2-cyclopenten-1-one,3-ethyl-2-hydroxy	0.81%	/	2.43
11.93	1,4:3,6-Dianhydro-d-glucopyranose	3.83%	/	1.90
12.094	3,4-anhydro-d-galactosan	35.90%	/	1.15
12.612	2,3-anhydro-d-mannosan	5.57%	/	1.28
12.73	Melezitose	1.43%	/	1.55
	Only from Lignin			
7.4	phenol	/	1.47%	0.53
9.277	phenol,3-methyl	/	1.27%	0.54
9.469	Phenol,2-methoxy-	/	10.81%	0.50
10.971	Phenol,4-ethyl-	/	3.23%	0.23
11.156	phenol,2-methoxy-4-methyl-	/	15%	0.36
12.002	Benzofuran,2,3-dihydro-	/	3.58%	0.89
12.98	phenol,4-ethyl-2-methoxy-	/	9.45%	0.26
13.611	2-methoxy-4-vinylphenol	/	6.83%	0.61
14.253	phenol,2,6-dimethoxy-	/	4.54%	0.47
14.336	phenol,2-methoxy-3-(2-propenyl)-	/	0.95%	0.50
14.495	phenol,2-methoxy-4-propyl	/	1.78%	0.25
14.981	phenol,2-methoxy-5-(1-propenyl)-	/	1.97%	0.37
15.97	phenol,2-methyoxy-4-(1-propenyl)-	/	6.99%	0.29
16.551	Acetoguaiacon	/	1.98%	0.45
17.086	5-tert-butylpyrogallol	/	1.83%	0.21
17.242	Guaiacylacetone	/	1.06%	0.36
17.719	3-tert-butyl-4-hydroxyanisole	/	1.24%	0.87
19.17	ethyl citrate	/	1.42%	0.40
19.807	Methyoxyeugenol	/	1.34%	0.38
	Others			
1.432	CO2	10.86%	4.14%	1.16

[a] LG-L is denoted for Levoglucosan-lignin

Title Mesoporous molecular sieve with strong acidity as catalysts support for pyrolysis of furfural residue to levulinic acid

Abstract Mesoporous molecular sieve with strong acidity has been prepared during a simple one step synthesis procedure. The resulting material was characterized by X-ray diffraction (XRD), transmission electronmicroscopy (TEM), N_2 adsorption/desorption, pyridine adsorbed FTIR spectroscopy and ^{27}Al MAS NMR techniques. The resulting material showed disordered worm-like mesoporous channels and strong acidity. Three TiO_2/ $SO4^{2-}$soliod catalysts were synthesized using ZSM-5□SBA-15 and self-made disordered worm-like molecular sieve as support respectively, and the catalytic performance of the resulting catalysts was examined in pyrolysis of furfural residue to levulinic acid(LA). The catalyst synthesised using self-made disordered worm-like mesoporous molecular sieve as support showed higher conversion of cellulose to levulinic acid(LA) than those of traditional microporous or mesoporous molecular sieve.

Presenter: Haiyan Xu, Energy Research Institute, Henan Academy of Sciences, and Henan Key Lab of Biomass Energy, China

Ethanol Production with Process of Simultaneous Saccharification and Fermentation at High Temperature

Xu Jingliang, Gao Yueshu, Liu Yunyun, Zhang Yu, Yuan Zhenhong, Zhuang Xinshu
Key Laboratory of Renewable Energy, Guangzhou Institute of Energy Conversion, Chinese Academy of Sciences, Guangzhou 510640, P.R. China
xjl@ms.giec.ac.cn
http://people.ucas.ac.cn/~xjl?language=en

Abstract: Simultaneous saccharification and fermentation (SSF) at high temperature is an attractive process for the cellulosic ethanol production. In the present study, different pretreatment processes, including liquid hot water (LHW) pretreatment, sodium hydroxide (NaOH) pretreatment, and their combined pretreatments, were conducted to improve the enzymatic digestibility and sugar recovery with sugarcane bagasse (SCB). LHW pretreatment could solubilize over 82.35% of xylan and 42.31% of lignin, and the NaOH pretreatment could remove 77.85% of lignin and retain most of glucan and xylan. For combined pretreatments, the sequence of two procedures had significant effects on the chemical composition, substrate characteristic and the subsequent enzymatic hydrolysis process. LHW-NaOH pretreatment could solubilize over 92.22% of xylan and remove 76.47% of lignin, and the highest total sugar recovery of 73.3% was achieved after 72 h of enzymatic hydrolysis. While NaOH-LHW pretreatment, which could remove nearly 84.33% of lignin, but only solubilize 71.15% of xylan, showed the highest enzymatic digestibility. The pretreatment efficiency order was as the following: NaOH-LHW > NaOH > LHW-NaOH > LHW. Taking the economy efficiency and simplicity into consideration, NaOH pretreatment was chosen as substrate pretreatment methods for the subsequent enzymatic hydrolysis and fermentation experiment.

To obtain high sugar yield and ethanol concentration, a fed-batch process was performed. Two processes namely simultaneous saccharification and fermentation (SSF) and separate hydrolysis and fermentation (SHF) for ethanol production had also been compared with commonly used *S. cerevisiae*. The results indicated that the ethanol concentration and yield in the SHF were higher, while the ethanol productivity (gram per unit volume and time) was lower than that of SSF. For 25% of substrate loading, the ethanol productivity and ethanol concentration could attain to 0.37 $g.L^{-1}.h^{-1}$, and 36.25 g/L for SSF in 48 hours of fermentation, respectively, while that for SHF process could attain to 0.32 $g.L^{-1}.h^{-1}$ and 47.95 g/L for SHF in 32 hours of fermentation. To alleviate the temperature incoordination during SSF process, a thermo-tolerant yeast, *Kluyveromyces marxianus* NCYC 587 was also employed, compared with *S. cerevisiae*. The results showed that the ethanol concentration remarkably increased to 42.21 g/L after 48 hours of fermentation and the ethanol productivity increased up to 0.44 $g.L^{-1}.h^{-1}$ with *Kluyveromyces marxianus*, and the maximum ethanol concentration, productivity, and yield (g ethanol /g cellulose) of SSF process with *Kluyveromyces marxianus* were all better than that of SSF process with *S. cerevisiae*. The studies also found that the ethanol concentration can improve by 30% with adding of β-glucosidase.

Key Words: Ethanol; Sugarcane bagasse; Pretreatment; Simultaneous saccharification and fermentation; Separate hydrolysis and fermentation;

Mechanism study of biomass pyrolysis

Bin Ru, Haizhou Lin, Qinjie Cai, Wuxing Sun, Shurong Wang*, Zhongyang Luo

State Key Laboratory of Clean Energy Utilization, Zhejiang University, Hangzhou310027, China

Biomass pyrolysis technology for liquid bio-fuels production has been developed for more than thirty years, while its industrial application is still restricted by the low grade of crude bio-oil. The mechanism study of biomass pyrolysis would contribute to improving the biomass pyrolysis technology, and providing guidance for bio-oil upgrading and reactor design. However, the structure of biomass is very complex, it is difficult to get insights into pyrolysis mechanism of the integral biomass. Since biomass is mainly composed of cellulose, hemicellulose and lignin, it is feasible to firstly investigate the pyrolysis mechanism of each individual components, followed by a modification of components interaction for the final integral biomass pyrolysis model. Cellulose, a polymer of glucose connected with β-(1,4)-glucosidic bond, usually has a maximum weight loss at around 350 °C. By using a high-radiation flash pyrolysis reactor, it is the first time for quantified characterization of the active cellulose, a key intermediate product of the Broido-Shafizadeh cellulose pyrolysis model. It is found that the active cellulose is a yellow, water-soluble solid material, which mainly composed of oligosaccharides. Additionally, density functional study (DFT) was introduced to investigate the possible pathway of cellulose decomposition with glucose as model compound. The Mulliken value analysis indicated that the glucose pyranoid ring could be easily opened. The followed transition state search proved that the formation of 5-hydroxymethyl-2-furaldehyde (HMF) with low energy barriers is the most favorable pathway for glucose pyrolysis. Hemicellulose, composed of many different monosaccharides, is another important carbohydrate in biomass. Xylan and other four hemicellulose monosaccharides, including two pentoses (xylose and arabinose) and two hexoses (mannose and galactose), were selected as model compounds. The TG-FTIR analysis indicated that the monosaccharides underwent similar degradation routes involving ring opening and carbon chain breakage, while xylan exhibited different decomposition behavior for the formation of acetic acid, CO_2, and CO due to the *O*-acetyl groups and 4-*O*-methylglucuronic acid units connected to the backbone. The followed DFT study indicated that furfural formation is more favorable than 1-hydroxy-2-propanone and 4-hydroxydihydrofuran-2(3H)-one during xylose pyrolysis. While hexoses preferred to form HMF because of the high energy barrier of the dissociation of hydroxymethyl group. And the breakage of O-acetyl groups for acetic acid formation could occurred at low temperature due to its lower energy barrier compared with that of the monosaccharides decomposition. As for lignin, a complex polymer of aromatic alcohols, is responsible for the phenols and pyrolytic lignin in the resulting bio-oil. FTIR, ^{1}H-NMR, ^{13}C-NMR and GPC were used to characterize the structures of lignin, it has high contents of methoxy groups and ether bond linkages. An introduced double-Gaussian distributed activation energy model indicated that the apparent activation energy for lignin pyrolysis is 148.77 kJ/mol. Moreover, it is found that the lignin from hardwood has higher content of syringyl units than softwood lignin, which results in the different distribution of pyrolytic products. And the characterization of pyrolytic lignin deduced that during the pyrolysis, the β-O-4 bonds in lignin are almost cleavage, yielding many aromatic oligomers, such as trimers and tetramers.

Acknowledgements: The authors are grateful for the financial support from the National Natural Science Foundation (51276166), the National Basic Research Program of China (2013CB228101).

SYNTHESIS OF POLYLACTIDE-GRAFT-MALEIC ANHYDRIDE VIA FREE RADICAL INITIATION IN SUPERCRITICAL CARBON DIOXIDE

Barbara A. Wheelden, Amber R. Tupper, Sunggyu Lee
Ohio University, Athens, OH

One of the most popular bio-based polymers on the market today is polylactide (PLA). Although PLA is bio-based, biodegradable, produced on a large scale, and able to be processed by traditional polymer processing equipment, its widespread use is limited by its relatively poor mechanical properties. PLA possesses a high tensile modulus, but low elongation and toughness. It is an inherently brittle material and therefore a poor choice for flexible packaging applications in its virgin form. PLA can be melt blended with more flexible polymers to increase its suitability for a wider range of applications, but due to their high molecular weight and viscosity, polymers are rarely miscible with each other. Immiscible polymer blends cause premature separation of polymeric phases, discontinuity of mechanical and thermal properties, and shortened service life. Graft copolymers can serve as compatibilizers between immiscible polymer blends to improve their miscibility and subsequent mechanical properties.

In this work, polylactide-graft-maleic anhydride (PLA-g-MA) will be synthesized with the ultimate objective of incorporating this graft copolymer as a compatibilzer into bio-based polymer blends with PLA as the main phase. PLA-g-MA will be synthesized in a stainless steel batch reactor in the presence of supercritical carbon dioxide ($scCO_2$). ScCO2 is an effective medium for polymerization because while most polymers do not dissolve, they tend to swell in scCO2, allowing for diffusion of reactive monomer into the swollen polymer. In addition, the density of the CO2 can be easily tuned by changing pressure and/or temperature, and the reaction can take place under ideal conditions without the harsh organic solvents utilized in solution grafting or high temperatures employed in molten state grafting.

In this study, the amount of initiator and MA will be kept constant at 2.5 wt% and 10 wt%, respectively, and the scCO2 density will be held at 300 kg/m^3. Azobisisobutyronitrile (AIBN) and Benzoyl Peroxide (BPO) will be used as the free radical initiators, and reactions will be carried out at three different temperatures for each of two initiators studied.

After the grafting reaction has taken place, the samples will be purified using Soxhlet extraction in methanol to remove excess monomer and spent initiator and Fourier Transform Infrared (FTIR) and Proton (^1H) Nuclear Magnetic Resonance (NMR) spectroscopy will be utilized to ensure that grafting has taken place and to determine the grafting level. The optimum reaction temperature and initiator will be determined based on these results.

387512 Ductility Improvement of Film Products Made of Nano-Fibrillated Saccharification Residues of Wood

Abstract:

The objective of this study was to investigate the effects of polymer incorporation on the tensile properties of films made of nano-fibrillated saccharification residues of wood. In earlier work, we demonstrated that it is technically feasible to utilize these lignin-rich residues for making plastic-like films, but the resulted films are brittle. In this study, we addressed the issue by incorporating a ductile thermoplastic at relatively low mass fractions, in contrast to the conventional approach (low fiber content) with nanofiber-polymer matrix composites. To implement this research, aspen wood flour was first pre-treated with an alkaline solution and then enzymatically saccharified to attain 80% glucose release from glucan. The solid residuals were mechanically fibrillated in aqueous media to fibrous fragments of micron and nanometer size. Some pre-treated wood flour (without saccharification) was also subjected to mechanical fibrillation. Fibrillated fragments from both pre-treated and saccharified biomass were blended at a solid mass ratio of 50:50. An appropriate amount of poly (vinyl alcohol) (PVA) solution was added to the mixture, and the slurry was immediately decanted to form into mats and subsequently dried to films. Results showed that PVA incorporation at 1.2% and 21% mass fraction of the films slightly improved tensile strain at break, but the films remained brittle. Increasing the PVA content to 37% mass fraction increased the strain to three folds, without compromising tensile strength of the films. From 24h-soaking study, incorporation of PVA also has the advantage of maintaining the film integrity in water with negligible weight loss (~1.0%). Other results such as wet strength and moisture adsorption will be discussed in the light of application potential. Overall, we successfully improved ductility while sustaining the tensile strength of nanofiber films. This outcome contrasts with the use plasticizers which usually increase ductility at the expense of strength (in addition to stiffness). By improving ductility of films containing saccharification residues, this study contributes to expanding the application window and value of the film products, thereby advancing the possibility of coproducing sugars and value-added materials from bioconversion of woody biomass.

Chemistry and chemical engineering process for making PET from bio based monomers.

Damian Adrian Salazar Hernandez

Advisor: Dr. Saleh A. Jabarin

Polyesters are one of the most important and versatile polymers in use today. They are produced by the polycondensation reaction of a glycol or dialcohol with a difunctional carboxylic acid. Polyesters are continuously used for several purposes ranging from bottles for water, carbonated drinks, health care products to fibers or carpet.

Poly ethylene terephthalate (PET) is polyester formed by step-growth polymerization from ethylene glycol (EG) and terephthalic acid (PTA). EG is traditionally obtained from the oxidation of ethylene over metal catalyst. PTA is obtained from liquid phase oxidation of p-xylene. The synthesis of PET requires a two stage process: esterification and polycondensation. PET raw material EG and PTA are currently obtained through reforming process of oil and natural gas. There are alternative chemical routes identified on how to successfully transform the sugars obtained from biomass material into terephthalic acid and ethylene glycol. Yield of final product obtained is still in the range of low molecular weight material and effective separation represents a challenging step for researchers. In this project we produce PET using EG and PTA obtained from oil and also with both raw materials obtained from a bio based source. Polymer produced with each set of raw material is characterized and compared to identify similarities or differences between them. In our approach we separate PTA precursor (p-xylene) from a bio-based aromatic mixture. This mixture can come from any of the following chemical routes: 1) bio based oil (BTX) provided to us, 2) product from Diels Alder reaction of 2,5-dimethylfuran (DMF) and ethylene, 3) product effluent from pyrolysis reaction of 2-methylfuran (2MF) and propylene. It is anticipated that in any of the separations there might be impurities (other organic compounds) that may affect the purity of the final product obtained; therefore our main objectives of the project are to determine the effect of these low molecular weight compounds (impurities) on the polymerization and characteristics of PET and identify critical areas on the separation train of precursor.

Thermochemical transformation of biomass produces furanic compounds by dehydration of sugar molecules present in the cellulose and hemicellulose. These compounds can be reacted with olefins via Diels-Alder reaction to produce aromatics. Using CDS Analytical Pyroprobe 2-methylfuran and propylene react at 600 °C over zeolite catalyst. Product obtained contains p-xylene along with other high carbon aromatics. This experiment was done to prove the concept on the formation of desired chemicals. The next step is to do pyrolysis in a fixed bed reactor to collect the product effluent and do separation of p-xylene. As extension of the work with furanic compounds, 2-5 dimethylfuran is reacted with ethylene over zeolite Y catalyst in a 4590 Micro Bench Top Parr Reactor. This reaction follows a cyclo addition mechanism producing aromatics with mono, di and tri substituted groups.

Bio oil and liquid samples obtained from separation train of experiments in the research are characterized using Shimadzu GC2010 with flame ionization detector, Bruker GC450 with mass spectrometer and Shimadzu VP HPLC with UV detector. In all cases PTA precursor is separated from each of the mixture using fractional distillation and crystallization. For the first step, we use a packed bed and a plate column along with a reflux distillation head to separate streams containing aromatics with different boiling point. The stream containing xylene isomers is cooled down to temperatures below -50 °C, this allows us to crystallize p-xylene and separate it via vacuum filtration. P-xylene, cobalt acetate, acetic acid and air are fed to Parr reactor to do oxidation. The product obtained is a low grade crude terephthalic acid that contains different impurities, i.e. 4-carboxylbenzaldehyde (4-CBA). The subsequent step is the purification, the crude acid is mixed with Pd over carbon and deionized water to create a slurry of 13% w acid. This mixture is placed in the same Parr reactor and hydrogen is used to pressurize the system. A hydrogenation reaction takes place, transforming the 4-CBA into toluic acid. The solid/liquid product is separated by vacuum filtration obtaining purified terephthalic acid in solid form. This bio based acid and bio based ethylene glycol (provided to us) are used to do polymerization on Parr reactor and produce PET. To make a comparison with traditional process, EG and PTA are obtained from industry and used to make polymer with the same procedure and reactor as the bio based.

Table 1 show the results obtained from analysis of bio oil. Benzene, toluene and xylenes account for 76% w of the total mixture, p-xylene in specific is present at 10% w. At the current stage of the research we are able to separate meta-, ortho- and para-xylene as a combined stream with 78% purity; we are working on crystallization to obtain p-xylene as a purified product.

Table 1 – Identified components in Bio oil.

A	Benzene	M	Benzene, 1,2,4,5-tetramethyl-
B	Toluene	N	1-propenal, 3-phenyl-
C	Ethylbenzene, m-p-xylene	O	Benzofuran, 2-methyl-
D	o-xylene	P	Benzene, 4-ethenyl-1,2-dimethyl-
E	Benzene-propyl	Q	2-methylindene
F	Benzene, 1-ethyl-3-methyl-	R	1H-Indene, 1-methyl-
G	Benzene, (1-methylethyl)-	S	Naphthalene
H	Benzene, 1,2,4-trimethyl-	T	Naphthalene, 1,2-dihydro-3-methyl
I	Benzene, 1,3,5-trimethyl-	U	Naphthalene, 1,2-dihydro-4-methyl
J	Indane	W	Naphthalene,2-methyl-
K	Indene	X	Naphthalene,2-ethyl-
L	Benzene, (2-methyl-1-1propenyl)-	Y	Naphthalene,1,2-dimethyl-

On the thermochemical approach, we followed a stage test to determine the effect of catalyst and propylene on the formation of aromatic material. 2MF is tested on pyroprobe unit; the effluent of the reaction is send to Bruker GCMS for characterization. Once the chromatogram is obtained, the area under the curve for the desired peaks is measured. Table 2 show the ratio of the area under the curve considering the pyrolysis of sole 2MF as the reference value. We can see an increase in the formation of aromatics by adding catalyst to the system. Then when we add propylene to the reaction media there is a dramatic increase in the formation of xylenes. This proves the idea that the formation of aromatics follows Diels Alder process and the feed of an olefin boost reaction rate. In each of the chemical routes we have promising results towards obtaining polymer precursor from bio derived source.

Table 2 - Area ratio for selected aromatics in 2MF pyrolysis.

	Catalyst Location			Area Ratio			
Material	Tube	Reactor	Gas	Benzene	Toluene	EB + m/p-xylene	o-xylene
2-MF	X		He	1.0	1.0	1.0	1.0
2-MF		X	He	8.1	7.5	13.4	12.9
2-MF		X	HePro	11.3	22.1	65.3	73.0

392458 Co-Injection Moulding Bioplastics: Investigating Procedure and Performance

Abstract:

To help combat environmental issues like climate change and landfill waste, bioplastics are being developed to provide renewable and/or biodegradable polymer options. Compostable, corn-based polylactic acid (PLA) is strong and stiff, but brittle. Polycaprolactone (PCL), a petroleum-based compostable polymer, is very tough but has poor strength. Combining these plastics by melt blending is an established method to create a new material that balances their complementary properties; however, alternative processing methods have the potential to be more successful. Co-injection moulding is a manufacturing technique that combines polymers in a layered structure; it can help to produce more aesthetically pleasing products, reduce processing steps, and may result in improved mechanical performance. Determining operating parameters for optimal co-injection conditions for PLA and PCL will be investigated and final products will be compared to similar melt blended samples. Emphasis will be placed on proper part formation and impact resistance in evaluating a successful co-injected material.

Acknowledgement: The financial support from the Natural Sciences and Engineering Research Council (NSERC), Canada for the Discovery grants individual (to Mohanty) and the Ontario Ministry of Agriculture & Food and Ontario Ministry of Rural Affairs (OMAF-MRA) to carry out this research is gratefully acknowledged.

Ecological Glue – Towards a Sustainable Industry?

Merten Morales[a,*], Reto Bättig[a], Stavros Papadokonstantakis[a], Konrad Hungerbühler[a]

[a]Institute for Chemical and Bioengineering, ETH Zurich, CH-8093 Zurich, Switzerland

merten.morales@chem.ethz.ch

1. Background + Aims

The introduction of a new generation of biorefineries into the market presents a multi complex problem, where questions like environmental sustainability, economic feasibility and conceptual exploration need to be answered beforehand. The biorefineries of the second generation need to tackle the objective of how chemicals can be produced from biomass, and not compete with the food production. Lignocellulosic biomass presents a raw material with a high potential to replace corn or sugarcane. Innovative production routes from wood to chemicals have to be included within the biorefinery concept. The quest of finding the optimal biorefinery concept is becoming even more complex by considering besides the conventional also innovative processes for the production of chemicals. The approach for finding the optimal concept is that generic optimization algorithm, developed by EPFL in Lausanne, is able to integrate different processes in order to find a bio refinery concept under the objective of profit maximization and environmental impact minimization.
A major part of the research project is the modelling of the different production paths to produce either platform chemicals (e.g. Glucose, Ethylene, Succinic acid) or high value added chemicals (Polyurethanes, Polyesters, Vinyl polymers). For the production routes conventional concepts as well as innovative concepts are considered for finding the optimal production routes. The concepts are evaluated through environmental indicators (Global Warming Potential, Ecoindicator 99, Cumulative Energy Demand) and economic indicators (Profit, Operational costs, Investment cost).

2. Methods + Results

An important platform chemical, which needs to be considered for a biorefinery, is acetic acid. The main application of acetic acid is the production of vinyl acetate, which is used to produce polyvinyl acetate (PVA). PVA is a common chemical for glue or latex emulsions (e.g. application paper coating, textile treatment) [1]. Nowadays acetic acid is mainly produced from petrochemical resources. The most important synthetic route over the last 25 years to produce acetic acid is the carbonylation of methanol. For introducing a feasible route for acetic acid in a biorefinery concept we modelled and optimized the process of Wacker Chemie [2]. The Wacker process is converting ethanol to acetic acid by the gas phase oxidation of ethanol. The major aim of this research project is to point out the major environmental and economic cost in the production of PVA from lingocellulosic biomass. The results of the project show the economic and environmental impact of different steps for the production of PVA.

1. Converting Wood to Glucose
2. Fermentation of Glucose to Ethanol
3. Dehydration of Ethanol to Ethylene
4. Oxidation of Ethanol to Acetic acid
5. Addition of Ethanol and Acetic acid to Vinyl acetate
6. Polymerization of Vinyl acetate to PVA

3. Summary/Conclusion

Through the assessment based on different rigorous models a conclusion can be drawn if PVA produced from the renewable resource wood can be produced more environmental friendly or is more economical attractive than PVA coming petrochemical industry.

References

[1] Ullmann`s Enycylopedia of Industrial Chemistry, Acetic acid, ISBN: 9783527306732

[2] Wacker Chemie AG, Method for producing Acetic and Formic acid by the gas phase oxidation of ethanol, WO 2008/110468 A1

366034 Chitosan from Solid-State Fermentation of Soybean Residues By Filamentous Fungi As Biobased Paperboard Coating Additives for HVAC Applications

Abstract:

The HVAC (Heating, Ventilation, and Air-conditioning) industry markets phenolic resin coated-paperboard media for use in the air-cooling of agricultural livestock pens and enclosures. Phenolic resins such as phenol formaldehyde have been shown to improve dry stiffness and water resistance of paper base sheets. However, in HVAC applications where good wet strength and water absorption properties are required, phenol formaldehyde-coated paperboards unfortunately have reduced wet stiffness levels compared with uncoated paperboards, thus resulting into premature failure of the air-cooling media. Furthermore, the use of phenolic resins may reduce the recyclability or biodegradability of the used coated paperboards. Chitosan, a versatile biopolymer commonly derived from crustacean shells in seafood wastes, has found numerous applications across multiple disciplines due to its unique and interesting properties. Some of these properties include its antimicrobial quality, chelating properties, and the ability to form hydrogels, fibers, and films. Chitosan has also found potential applications in the pulp and paper industry, mainly as surface coating or sizing additive for paper and paperboard products with enhanced or functionalized properties. These properties include water resistance, air barrier, wet and/or dry strength, and antimicrobial properties. Thus, chitosan could serve as a bio-based alternative to conventional phenolic resin as paperboard coating materials. However, the current process for chitosan production involves costly, labor-intensive, and harsh chemical treatments of seafood waste and also suffers from raw material supply inconsistencies and geographical limitation. Fortunately, chitosan can also be found as cell wall components of filamentous fungi. These fungi can be grown in fermentation cultures using agricultural and food processing residues as substrates to produce high-quality chitosan continuously and sustainably, and with expectedly lower separation and recovery costs.

The current study investigated the use of soybean processing residues such as meal and hulls as substrate for the growth and chitosan production of fungi using solid-state fermentation (SSF). Selected fungal strains with documented high chitosan production capabilities were screened for direct growth and utilization of solid soybean residue fermentation substrate. Results show that three out of the six screened fungi strains exhibited expansive mycelial growth on soymeal substrate while the rest only showed liquefaction of the soymeal substrate. The fungal strains exhibiting growth on the soymeal substrate were selected for fermentation kinetics and optimization studies to determine optimal SSF conditions for growth and chitosan production. The product chitosan is intended for application as coating additives for paper-based evaporative cooling media requiring wet stiffness and water absorption/wicking characteristics. Economic analysis indicate a target chitosan yield of 20 % (w/w of substrate) in order to match the price of the conventional phenolic resin coatings of $3.00 per lb. Further cost reductions are possible with lower chitosan application weights or coating techniques for optimum wet stiffness and wicking performance. Parallel investigations were conducted using commercially available chitosan to determine the optimum levels of chitosan concentration in carrier solution, coating weight, and other coating process variables for improved coated paperboard dry and wet stiffness and water absorption/wicking. Chitosan-coated paper basesheets, along with the uncoated and phenolic resin-coated sheets were characterized and compared for Wet Taber Stiffness and Wicking Height according to the manufacturer's methods and specifications. The target critical-to-quality (CTQ) parameters and their levels for application as cooling pads are 1600 mg wet Taber stiffness and 2 inches of wicking height over 10 minutes. The economics of paperboard coating using chitosan compared with phenolic resins based on coating amounts on weight or area basis, unit costs, and product performance will be further discussed.

393363 R&D of Green and High Efficient Cellulosic Fuel Ethanol Process Combined with Value Added By-Products By Three Main Feedstock Components

Abstract:

As the biggest developing country, China needs fuel ethanol to meet its demands of the social and economy development. Basing on the "Mid-long term plan for renewable energy" established in 2007, the annual consumption of fuel ethanol will reach 10 million tons in 2020. Cellulosic ethanol, one of the most important second generation biofuel, gains more and more attentions for its substitution of the transportation liquid, lessen the air pollution by the vehicles and promote the rural area economy. In Guangzhou Institute of Energy Conversion, a green and high efficient Cellulosic fuel ethanol process combined with value added by-products by three main feedstock components is proposed. One ton ethanol can be produced with 5.68 t sugarcane bagasse, as well as 0.96 t CO2, 0.96 t Xylooligosaccharides and 1.59 t lignin. The lignin was used for the mushroom medium which can produce nutrient - rich *Lentinula edodes* successfully.

378318 Effect of Phospholipids on Free Lipase-Catalyzed Biodiesel Production

Abstract:

Free lipase-catalyzed biodiesel production has drawn more and more attentions in recent years because of its advantages of lower cost and faster reaction rate. Utilizing free lipase to convert low quality oils such as crude vegetable oils and microbial oils is beneficial to further reduce the cost of biodiesel production. However, these oils typically contain some amount of phospholipids. Phospholipids were found to affect the lipase-catalyzed process and further influence the enzyme's thermal stability in biodiesel production process. In this work, free lipase-catalyed biodiesel production was systematically investigated with oils containing phospholipids as feedstocks at varied conditions. It was found that increasing temperature led to a decreased fatty acid methyl esters (FAME) yield and poor reuse stability of the lipase in converting oils containing phospholipids for biodiesel production. Further study showed that this inhibitory effect was mainly attributed to the coexistence of phospholipids and methanol in the system. A novel two-step enzyme-catalyzed process was developed, with which the above-mentioned inhibitory effect was eliminated, and a FAME yield of 95.1% could be obtained with oils containing 10% phospholipids as the feedstock.

Study on syngas production by catalytic gasification of maize stalk fast pyrolysis bio-oil

Weiming Yi, Zhihe Li, Peng Fu, Xueyuan Bai, Nana Wang, Hongzhen Cai, Lihong Wang

School of Agricultural and Food Engineering, Shandong University of Technology, Zibo 255049, China
*Corresponding author. Tel.: +86 533 2786308; Fax: +86 533 2786158

Abstract

Hydrogen production via catalytic gasification of maize stalk fast pyrolysis bio-oil over the nickel/alumina supported catalysts promoted cerium was studied using a laboratory scale fixed bed coupled with Fourier transform infrared spectroscopy/thermal conductivity detection analysis (FTIR/TCD). The effects of nickel loading, reaction temperature, steam to carbon molar ratio (S/C) and bio-oil weight hourly space velocity (W_bHSV) on syngas production were investigated.

The results showed that increasing reaction temperature from 600 to 900 °C resulted in the significant increase of hydrogen yield and carbon monoxide yield. Meanwhile, the carbon conversion increased, but its increment was less pronounced than that of hydrogen yield. It was noteworthy that the carbon conversion did not reach 100% even at 950 °C. The difference related to the carbonaceous materials deposited on the catalyst bed, the carbon-containing products such as C_3H_8, C_3H_6, C_4H_{10} etc. that were difficult to be determined and the unconverted reactant itself. The hydrogen yield was significantly enhanced by increasing the S/C ratio from 1 to 3, whereas it increased slightly by further increasing S/C ratio. The CO yield always deceased with increasing the S/C ratio. The opposite trend was observed for CO_2. This was mainly because increasing the S/C ratio is conducive to the proceeding of water gas shift reaction towards hydrogen production, leading to the continuous decrease of CO yield and the increase of H_2 and CO_2 yields. The hydrogen yield decreased with the increase of W_bHSV. Meanwhile, the coke deposition amount changed little with increasing W_bHSV up to 12 h^{-1} and then it increased by 62.9% with the further increase of W_bHSV.

***Keywords*:** Maize stalk; Bio-oil; Catalytic gasification; Syngas production

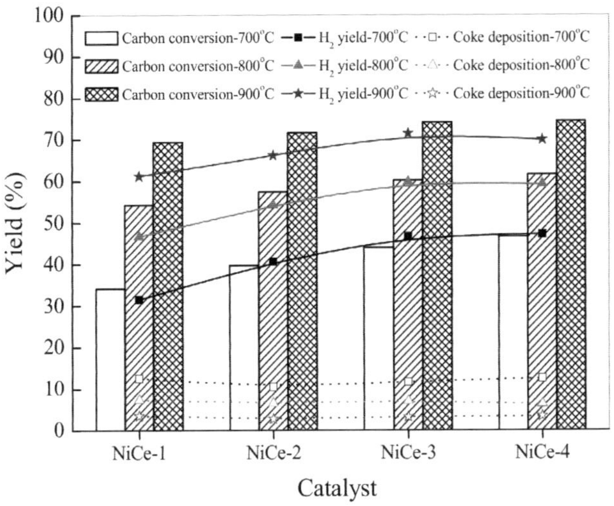

Fig. 1. Comparison of the activity of the four Ni-Ce/Al_2O_3 catalysts with different Ni loadings at the operational conditions of WCMR = 6 and W_bHSV = 12 h^{-1}.

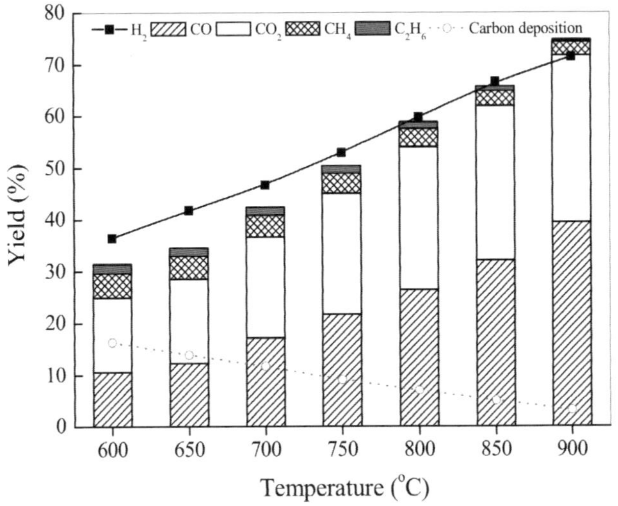

Fig. 2. Effect of reforming temperature on the yields of hydrogen and the main carbon-containing gaseous products as well as coke deposition percentage over NiCe-3 at the operational conditions of WCMR = 6 and W_bHSV = 12 h^{-1}.

Title Support effect on Fischer-Tropsch synthesis of gas from biomass gasification over a Ru-Cu-Co/Al_2O_3 catalyst

Abstract The Al_2O_3-supported Ru-Cu-Co catalysts were modified by loading with several additives (Mg Zr, and La) by incipient wetness impregnation and tested in the Fischer Tropsch reaction using a simulated syngas (CO/H_2) mixture similar to that obtained in biomass gasification. The catalysts characterization included N_2 adsorption, X-ray diffraction (XRD), temperature programmed reduction (TPR) and scanning electron microscopy (SEM) techniques. The modification of support has a great influence on the interaction between active metal and support and then the reducibility of active metal oxides.The result shows that the additives affected the catalyst activity and the C_{5+} selectivity and Zr was the best species for improving the performance.

Presenter: Shunqing LI, Energy Research Institute, Henan Academy of Sciences, and Henan Key Lab of Biomass Energy, China.

Co-cracking of bio-oil fraction and ethanol for aromatic hydrocarbon production

Qinjie Cai, Li Zhang, Junhao Chen, Bin Ru, Shurong Wang,* Zhongyang Luo

State Key Laboratory of Clean Energy Utilization, Zhejiang University, Hangzhou310027, China

Catalytic cracking using solid acidic catalyst like HZSM-5 is a promising way to transform the highly oxygenated biomass pyrolysis oil (bio-oil) into hydrocarbons, especially aromatic hydrocarbons. However, because the crude bio-oil has a high oxygen content and a high degree of unsaturation, corresponding to a low $(H/C)_{eff}$ ratio, coke easily forms during cracking, which leads to catalyst deactivation and reactor blockage. Using ethanol as the co-cracking reactant, which has a relatively high $(H/C)_{eff}$ ratio of 2, will increase the integral $(H/C)_{eff}$ ratio and thus suppress the coke formation. Moreover, some large molecular sugars and phenol oligomers in bio-oil are non-volatile and easily coke in the catalyst bed, and using molecular distillation can effective remove these compounds and enriched reactive acids and ketones in the distilled fraction (DF), which has a better cracking performance than crude bio-oil. Therefore, the bio-oil cracking technology is improved by using molecular distillation to separate bio-oil and obtain the DF first, and then adopting ethanol as the co-cracking reactant. Based on this technical route, the typical model compounds in the DF, namely hydroxypropanone, cyclopentanone, acetic acid, were selected to study their individual co-cracking performance with ethanol first. Under the condition of 400 °C and 2 MPa, with the model compound to ethanol weight ratio of 3:7, the reactants were converted completely over HZSM-5 catalyst. The yields of oil phases for co-cracking of hydroxypropanone, cyclopentanone, acetic acid with ethanol reached 31.9 wt%, 32.1 wt%, and 33.4 wt%, respectively. The aromatic hydrocarbon content in these oil phases was around 95%, and the rest were aliphatic hydrocarbons. Then, the actual bio-oil DF was used to be co-cracked with ethanol. The effect of reaction temperature, pressure, and the DF/ethanol ratio in the feed was investigated. It was found that both increasing the reaction temperature and adopting pressurized cracking benefited the yield and quality of oil phase. Using the optimum reaction temperature and pressure, co-cracking of the DF and ethanol, with different weight ratios, all generated high-quality oil phases. At 400 °C and 2 MPa, co-cracking of DF and ethanol with a weight ratio of 2:3 produced a high oil phase yield of 25.9 wt %. The aromatic hydrocarbon content in this oil phase was 94%. These generated aromatic hydrocarbons are not only important components in the commercial gasoline, but also promising chemical raw materials, especially the benzene, toluene and xylenes (BTX).

Acknowledgements: The authors are grateful for the financial support from the National Natural Science Foundation (51276166), the National Basic Research Program of China (2013CB228101).

Title Conversion of Corncob into Furfural by Acid and $FeCl_3$ Co-catalysis Process

Abstract Corncob was hydrolyzed by acid which is in steam state and $FeCl_3$ co-catalyzed process in a semi-batch tubing-bomb reactor system to produce furfural. Acid (such as acetic acid, formic acid and hydrochloric acid) and $FeCl_3$ co Catalyzed process can promote the furfural yield. A series of experiments were carried out to investigate the effects of acid, $FeCl_3$ concentrations and temperatures on furfural production. A maximum furfural yield of 72.89% (35.74% higher than that in conventional sulfuric acid-catalyzed process) was obtained at 180°C in the presence of 12g $FeCl_3$ and 3% of acid. Acid and $FeCl_3$ co-catalyzed hydrolysis was not only a high efficiency and environmental friendly technique, but also provided a possibility to utilize the furfural residue for ethanol production and other industries, such as ethanol production, levulinic acid production and so on.

Presenter: Suxia Ren, Energy Research Institute, Henan Academy of Sciences, and Henan Key Lab of Biomass Energy, China

Title Performance and emission characteristics of a diesel engine using alternative vehicle fuels formulations based on ethyl levulinate

Abstract In the paper, ethyl levulinate (EL) was taken as the key addition fuel, physical and chemical properties of different fuel formulas of EL-biodiesel-diesel were studied based on China National Standards and Codes: GB/T25199-2010 biodiesel fuel blend (B5). 13 kinds of fuel formulations were gotten and 3 of them were screened. According to theories of grey relation analysis(GRA) and analytical hierarchy process(AHP), synthesizes evaluation methods to evaluate fuel formulas was established with considering 16 kinds of indexes. The fuel formulas that satisfying the B5 standards were optimized and 4 kinds of the optimized fuel formulas were gotten. Optimized fuel formulas were selected to evaluate engine performance and emissions using an unmodification diesel engine test bench. The results show that engine power and torque with the fuel formulas were in general similar to those with diesel. Both brake specific fuel and energy consumption were lower with fuel formulas than with diesel. In general, the emissions of HC and CO and smoke opacity reduced significantly with fuel formulas compared with diesel while the emissions of NO_x and CO_2 increased.

Presenter: Zhiwei WANG, Energy Research Institute, Henan Academy of Sciences, and Henan Key Lab of Biomass Energy, China.

AUTHOR INDEX

AUTHOR INDEX

AUTHOR INDEX

AUTHOR INDEX

AIChE
120 Wall Street, FL 23
New York, NY 10005-4020

ISBN 978-1-5108-1260-4

CURRAN ASSOCIATES INC.
proceedings.com